BLAKE'S

Maths Guide

for Lower primary students

Bev Dunbar

Blake's Maths Guide
Lower Primary

ISBN: 978 1 74215 941 6

Published by Pascal Press
PO Box 250
Glebe NSW 2037
www.pascalpress.com.au
contact@pascalpress.com.au

Author: Bev Dunbar
Publisher: Lynn Dickinson
Editor: Tim Learner
Proofreader: Vanessa Barker
Design and illustration: Janice Bowles
Typeset by Julian Mole, Post Pre-press Group
Printed by Green Giant Press

CONTENTS

AUSTRALIAN CURRICULUM CORRELATIONS – YEAR 1

NUMBER and ALGEBRA	ELABORATIONS	ACMNA	PAGE
Number and place value Develop confidence with number sequences to and from 100 by 1s from any starting point. Skip count by 2s, 5s and 10s starting from 0.	★ Use skip counting games ★ Develop fluency counting forwards and backwards ★ Describe number sequences	12	1–6
Recognise, model, read, write and order numbers to at least 100. Locate these numbers on a number line.	★ Model using a wide range of materials ★ Identify numbers on a prepared number line	13	17–19
Count collections to 100 by partitioning numbers using place value.	★ Focus on groups of 10 and breaking 2-digit numbers into 10s and 1s	14	20–25
Represent, solve simple addition and subtraction problems using a range of strategies including counting on, partitioning and rearranging parts.	★ Develop a range of mental strategies for addition and subtraction problems	14	30–31
Fractions and decimals Recognise and describe one-half as one of two equal parts of a whole.	★ Share sets of objects into 2 equal portions ★ Split objects into 2 equal pieces ★ Describe how the pieces are equal	16	68–69
Money and financial mathematics Recognise, describe and order Australian coins according to their value.	★ Compare Australian coins to those in other countries ★ Understand the value of a coin is not related to size ★ Describe features to identify and distinguish coins	17	74–81
Patterns and algebra Investigate and describe number patterns formed by skip counting and patterns with objects.	★ Use place-value patterns to generalise and predict in a number sequence ★ Investigate patterns in the number system	18	87–90
MEASUREMENT and GEOMETRY		**ACMMG**	
Using units of measurement Measure and compare the lengths and capacities of pairs of objects using uniform informal units.	★ Understand that when you compare objects the unit must be the same size	19	93–99
Tell time to the half-hour.	★ Read time on analogue and digital clocks	20	122–123
Describe duration using months, weeks, days and hours.	★ Describe duration of familiar situations	21	124–128
Shape Recognise and classify familiar 2-dimensional shapes and 3-dimensional objects using obvious features.	★ Focus on geometric features ★ Use everyday words (corners, edges, faces) to describe shapes and objects	22	133–135
Location and transformation Give and follow directions to familiar locations.	★ Understand that directions involve turns, direction, distance ★ Understand direction words such as clockwise, anti-clockwise, forward, under ★ Interpret and follow directions around familiar locations	23	156

STATISTICS and PROBABILITY	ELABORATIONS	ACMSP	PAGE
Chance Identify outcomes of familiar events involving chance and describe them using everyday language such as 'will happen', 'won't happen' or 'might happen'.	★ Justify that some events are certain or impossible	24	160–161
Data representation and interpretation Choose simple questions and gather responses.	★ Determine which questions will gather appropriate responses for a simple investigation	262	164
Represent data with objects and drawings where one object or drawing represents one data value. Describe the displays.	★ Understand 1-1 correspondence ★ Describe displays by identifying those with the greatest or least number of objects	263	166–167

AUSTRALIAN CURRICULUM CORRELATIONS – YEAR 2

NUMBER and ALGEBRA	ELABORATIONS	ACMNA	PAGE
Number and place value Investigate number sequences, initially those increasing and decreasing by twos, threes, fives and tens from any starting point, then moving to other sequences.	★ Develop fluency and confidence with numbers and calculations ★ Recognise patterns in number sequences	26	7–14
Recognise, model, represent and order numbers to at least 1000.	★ Recognise there are different ways to represent numbers and identify patterns beyond 1000 ★ Develop fluency with writing numbers in meaningful contexts	27	15–16
Group, partition and rearrange collections up to 1000 in hundreds, tens and ones to facilitate more efficient counting.	★ Use an abacus to model numbers ★ Understand 3-digit numbers have 100s, 10s and 1s ★ Use a variety of place-value models	28	26–29
Explore the connection between addition and subtraction.	★ Become fluent with partitioning numbers ★ Use counting on to identify missing elements in an addition problem	29	32–38
Solve simple addition and subtraction problems using a range of efficient mental and written strategies.	★ Become fluent with mental strategies such as commutativity, building to 10, doubles, 10s facts and adding 10 ★ Model simple addition problems using 10s frames, 20s frames and empty number lines	30	40–51
Recognise and represent multiplication as repeated addition, groups and arrays.	★ Model arrays and explain reasoning ★ Visualise a group of objects as a unit and use this to calculate the number of objects in a larger group	31	52–61
Recognise and represent division as grouping into equal sets and solve simple problems using these representations.	★ Divide a collection into equal-size groups ★ Identify the difference between sharing (divide a set into 3 equal groups) and grouping (divide a set into groups of 3)	32	62–67
Fractions and decimals Recognise and interpret common uses of halves, quarters and eighths of shapes and collections.	★ Demonstrate different fractions by partitioning sets. Relate the number of equal parts to the size of a fraction	33	70–73
Money and financial mathematics Count and order small collections of Australian coins and notes according to their value.	★ Identify equivalent values in coins or notes. Count coins or notes to a given value	34	82–86

AUSTRALIAN CURRICULUM CORRELATIONS – YEAR 2 CONTINUED

NUMBER and ALGEBRA (continued)	ELABORATIONS	ACMNA	PAGE
Patterns and algebra Describe patterns with numbers and identify missing elements.	★ Describe skip counting patterns and represent the patterns on a number line ★ Investigate features of a pattern resulting from adding 2s, 5s or 10s	35	91
Solve problems by using number sentences for addition or subtraction.	★ Represent word problems as a number sentence ★ Write word problems to match a given number sentence	36	92
MEASUREMENT and GEOMETRY		**ACMMG**	
Units of measurement Compare and order several shapes and objects based on length, area, volume and capacity using appropriate uniform informal units.	★ Compare lengths using finger widths, hand span or a piece of string ★ Compare areas using your palm or a variety of flat informal units ★ Compare capacities of a range of containers	37	100–109
Compare masses of objects using balance scales.	★ Use balance scales to check if a mass is more than, less than or the same mass as another ★ Use informal masses to balance an object	38	110–121
Tell time to the quarter-hour, using the language of 'past' and 'to'.	★ Identify and describe the position of the hands on an analogue clock	39	125–127
Name and order months and seasons.	★ Compare ways different cultures identify the seasons ★ Recognise the seasons' connection to weather patterns	40	129–130
Use a calendar to identify the date and determine the number of days in each month.	★ Locate and identify specific information on a calendar	41	131–132
Shape Describe and draw 2-dimensional shapes, with and without digital technologies.	★ Identify key features of squares, rectangles, triangles, kites, rhombuses and circles, such as straight or curved lines, the number of edges or corners	42	143–154
Describe the features of 3-dimensional objects.	★ Identify geometric features such as faces, corners or edges	43	133–142
Location and transformation Interpret simple maps of familiar locations and identify the relative positions of key features.	★ Use representations of objects and their position to receive and give directions, describe places ★ Construct arrangements of objects from a set of directions	44	157–159
Investigate the effect of 1-step slides and flips with and without digital technologies.	★ Understand objects can be moved but changing position does not alter size or features	45	154
Identify and describe half and quarter turns.	★ Predict, reproduce a pattern based on 1/4 and 1/2 turns of a shape ★ Sketch the next element in a pattern	46	155

STATISTICS and PROBABILITY	ELABORATIONS	ACMSP	PAGE
Chance Identify practical activities and everyday events that involve chance. Describe outcomes as 'likely' or 'unlikely' and identify some events as 'certain' or 'impossible'.	★ Classify a list of everyday events using the language of chance ★ Explain your reasoning	47	162–163
Data representation and interpretation Identify a question of interest based on one categorical variable. Gather data relevant to the question.	★ Determine the variety of birdlife in the playground ★ Use a prepared table to record observations.	48	165
Collect, check and classify data.	★ Recognise usefulness of tally marks ★ Identify and use categories to sort data	49	168
Create displays of data using lists, table and picture graphs and interpret them.	★ Represent data with 1-1 picture graphs ★ Compare usefulness of different data displays	50	168–169

HOW TO USE THIS BOOK

Mathematics is a way of thinking. It helps you understand how the world works. *Blake's Maths Guide for Years 1 and 2* helps you see mathematics all around you. This Guide helps you to talk about, to draw and to record Mathematics. You will have the tools you need to be a successful mathematician.

The definitions are clear, concise and written in user-friendly language. Photographs show you how mathematics is used in our everyday world.

In the TRY THIS section, you practise, make or imagine mathematical concepts. This helps you put the ideas inside your head. Selected answers are at the back of the book.

Blake's Maths Guide for Years 1 and 2 contains an index to help locate the specific word you need more information about. The glossary at the back gives you quick definitions.

This Guide is a vital reference if you want to be successful at year 1 and 2 mathematics.

ABOUT THE AUTHOR

Bev Dunbar is a highly respected mathematics educator. Over the last 35 years, Bev has worked extensively with students, student teachers, parents and teachers within both government and non-government education systems. She has also lectured in Mathematics Education at the University of Sydney and the Australian Catholic University. Bev is the author of many educational resources, including *Times Tables 1* and *2*, 16 books for teachers in *The Exploring Maths* series and 10 books for students in the *Excel Maths Early Skill* series.

Bev is dedicated to helping you understand and enjoy mathematics. Her personal interests include a passion for painting and recreating medieval artworks and, of course, unravelling challenging sudokus.

COUNT TO 100

Let's find out how numbers work

NAME THE POSITION

You can put things in order when you count.
Position names are **ordinal numbers**.
Such as in a race where you come first, second or third.

The shark is on the left. It is first.

The turtle is fifth from the left.

The rabbit is tenth from the left.

Position names sound different to counting names.
Many have a "th" at the end.
You need to listen carefully.

1		first	1st
2		second	2nd
3		third	3rd
4		fourth	4th
5		fifth	5th

6		sixth	6th
7		seventh	7th
8		eighth	8th
9		ninth	9th
10		tenth	10th

Try this

Look at this line of people.

Starting from the right-hand side:

Put a △ under the 1st person.

Put an X under the eighth person.

Draw a line around the 11th person.

In what position is the girl in the brown suit? ____________

COUNT BY 1s TO 20

People love to count.

1, 2, 3, 4, 5, 6, 7, 8, 9,10,11,12,13,14

But it is hard to count messy things.

It is easy to count tidy things.

1 2 3 4 5

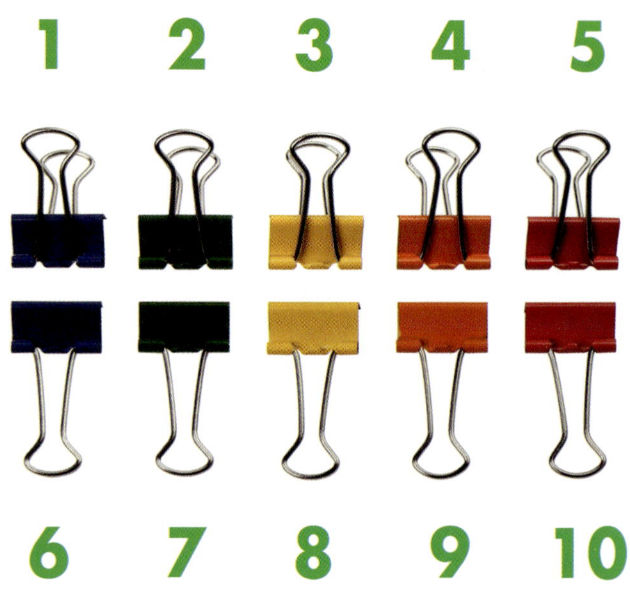

6 7 8 9 10

Try to touch each object as you count aloud.

Or draw a line or cross things out as you count.

1, 2, 3, 4, 5, 6, 7, 8, 9,10,11,12,13,14,15,16,17,18,19, 20

Counting backwards is the same order, just reversed. Try to see the numbers in your mind.

20 19 18 17 16 15 14 13 12 11 10 9 8 7 6 5 4 3 2 1

How many birds?

Try this

How many candles?

How many triangles?

How do you work it out?

COUNT BY 1s TO 100

Start from any number. Count forwards and backwards

When you count forwards by 1s the next number is one more. The numbers are getting larger.

When you count backwards by 1s the next number is one less. The numbers are getting smaller.

SKIP COUNT BY 2s TO 100

0 1 **2** 3 **4** 5 **6** 7 **8** 9 **10** 11 **12** 13 **14** 15 **16** 17 **18** 19 **20**

Here are 20 cats.

All the numbers have **0 2 4 6** or **8** in the ones digit.

It is a pattern of **even** numbers.

If there is 1 left over just count 1 more.

If you skip count by 2s, you only count every second object

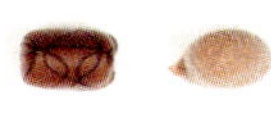

22 … **24** … **26** … **28** and 1 more makes **29**

There are 29 chocolates.

A box jellyfish has 24 eyes.

Start at any number when you count forwards by 2s.

53 55 57 59 61 63 65 67 79 71 73

All the numbers have **1 3 5 7** or **9** in the ones digit.

This is a pattern of **odd** numbers.

Start at any number when you count backwards by 2s.

88 86 84 82 80 78 76 74 72 70 68 **or**

65 63 61 59 57 55 53 51 49 47 45

Challenge

Close your eyes.

Start at 94. Count backwards by 2s to 70.

0 2 4 6 8 10 12 14 16 18 20 22 24 26 28 30 32 34 36 38 40 42 44 46 48 50 52 54 56 58 60 62 64 66 68 70 72 74 76 78 80 82 84 86 88 90 92 94 96 98 100

2 less than 94 (92)   2 more than 94 (96)

SKIP COUNT BY 5s TO 100

If you skip count by 5s, you only count every fifth object.

0 5 10 15 20 25 30 35 40 45

Here are 45 dogs.

What number pattern can you see?

All the numbers have a **0** or a **5** in the ones place.

Some things you count are in a mess.

Try to make groups of 5 then count any left overs by 1s

5 10 15 20 25 ... 26 27

I can see 27 tigers.

Start at any number when you count forwards by 5s.

7 12 17 22 27
32 37 42 47
52 57 62 67

What number pattern can you see?

All the numbers have a **7** or a **2** in the ones place.

Start at any number when you count backwards by 5s.

96 91 86 81 76 71 66 61 56 51 46 41 36 31 26 21

What number pattern can you see?

All the numbers have a **6** or a **1** in the ones place.

Flick your fingers as you skip count by 5s aloud.

Start at 65. How fast can you go?

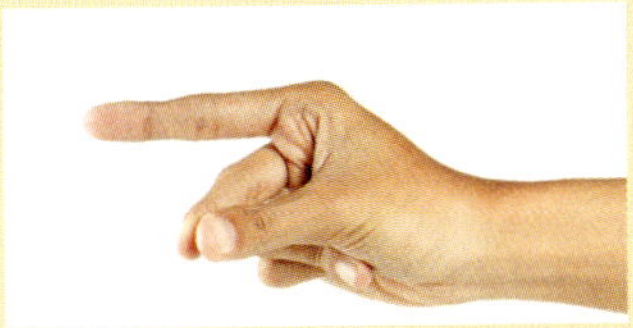

Try this

0 5 10 15 20 25 30 35 40 45 50

55 60 65 70 75 80 85 90 95 100

5 less than 80

5 more than 80

SKIP COUNT BY 10s TO 100

If you skip count by 10s, you only count every tenth object.

0 10 20 30 40 50 60

Here are 60 cars.

What number pattern can you see?

All the numbers have a 0 for the ones digit.

Some things you want to count are in a mess.

Try to make groups of 10 then count them.

Just count any left overs by 1s.

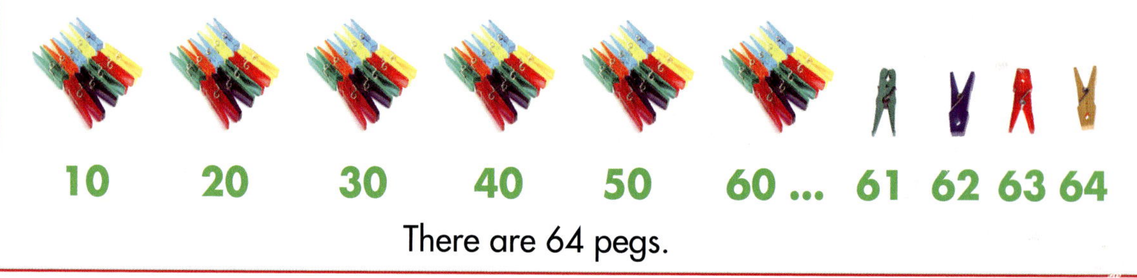

10 20 30 40 50 60 ... 61 62 63 64

There are 64 pegs.

Start at any number when you count forwards by 10s.

4 14 24 34 44 54 64 74 84 94

What number pattern can you see?

All the numbers have a **4** for the ones digit.

Start at any number when you count backwards by 10s.

96 86 76 66 56 46 36 26 16 6

What number pattern can you see?

All the numbers have a **6** for the ones digit.

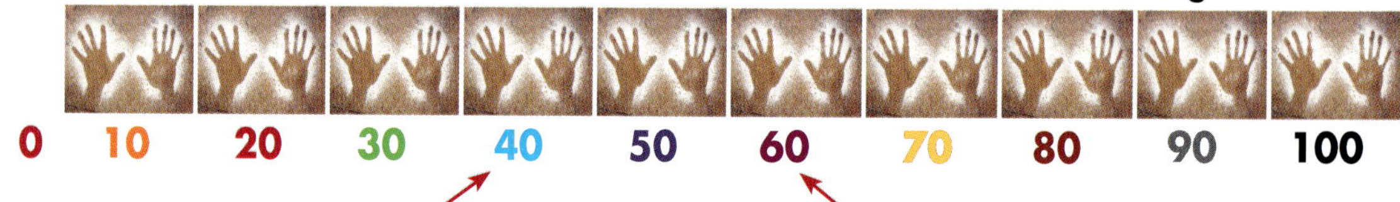

0 10 20 30 40 50 60 70 80 90 100

10 less than 50 10 more than 50

CALCULATOR COUNTS

Press **2 + +** then **=**

Keep pressing 

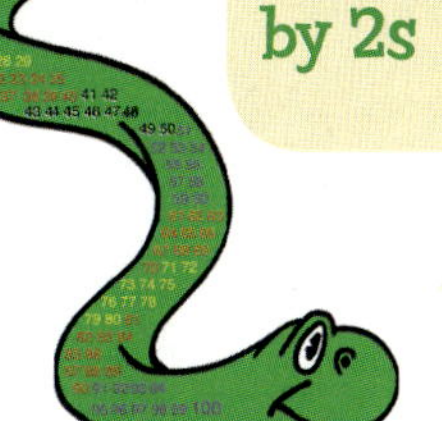

Skip count by 2s from any number on your calculator.

Press the start number then **+ 2 =**

59 + 2 =

Your calculator can skip count by 5s.

Press **5 + +** then **=**

Skip count by 5s from any number on your calculator.

Press the start number then **+ 5 =**

4 7 + 5 =

Your calculator can skip count by 10s.

Press **1 0 + +** then **=**

 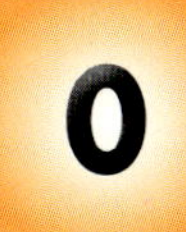   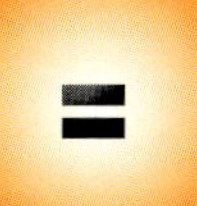

Skip count by 10s from any number on your calculator.

Press the start number then **+ 1 0 =**

2 3 + 1 0 =

    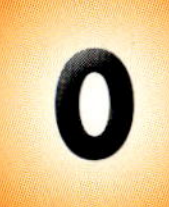

COUNT TO 100

0	1	2	3	4	5	6	7	8	9
10	11	12	13	14	15	16	17	18	19
20	21	22	23	24	25	26	27	28	29
30	31	32	33	34	35	36	37	38	39
40	41	42	43	44	45	46	47	48	49
50	51	52	53	54	55	56	57	58	59
60	61	62	63	64	65	66	67	68	69
70	71	72	73	74	75	76	77	78	79
80	81	82	83	84	85	86	87	88	89
90	91	92	93	94	95	96	97	98	99

Look at the grid.

What other number patterns can you see?

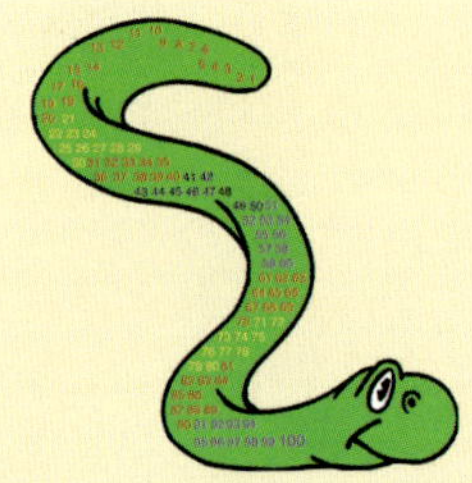

Challenge

Draw your own 10 × 10 square grid.

Write in the numbers from 0 to 99 without looking at the grid above.

ESTIMATE TO 100

Sometimes you estimate rather than count exactly.

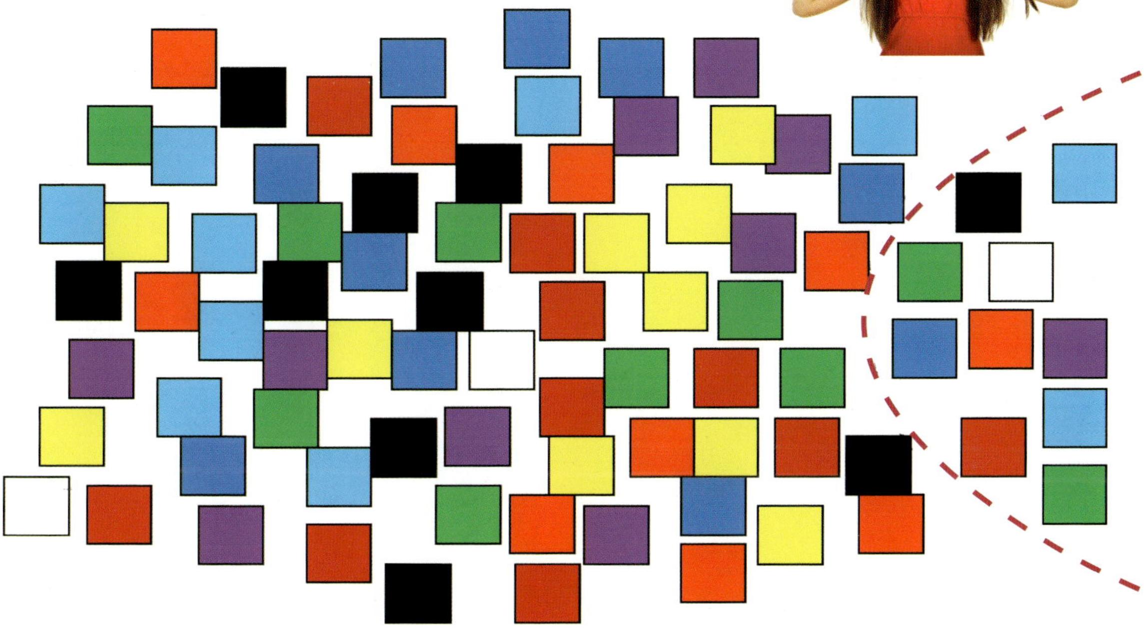

First count a small part of the whole group.

See what 10 looks like.

Imagine more groups this size.

How many groups like this will fit the whole group?

Count by 10 to find your estimate.

Jack Emmy Mario

There are 83 squares altogether. Mario's estimate was the closest to 83.

COUNT TO 1000

COUNT BY 1s TO 1000

You can count forever.

Try not to get mixed up when you reach the next 100.

Just count by 1s all over again to the next 100.

… 396 397 398 399 **400** 401 402 403 404

We counted 404 cars in 5 minutes.

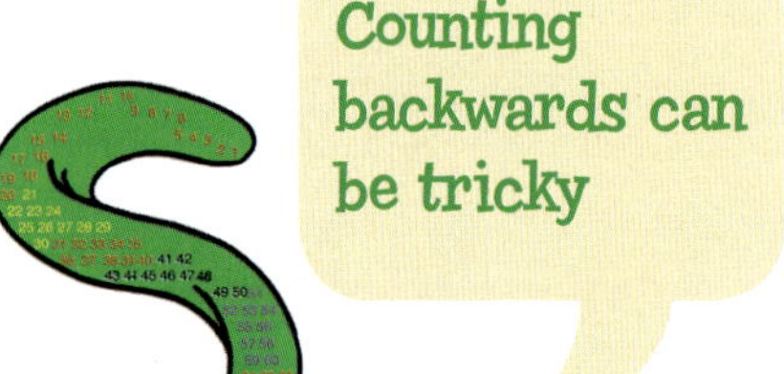

Try not to get mixed up when you go past another 100.

Try to see the numbers in your mind.

803 802 801 **800** 799 798 797 796

Count forwards by 1s from 362 to 380.

Count backwards by 1s from 711 to 688.

COUNT BY 2s TO 1000

You can count by 2s from any even number.

The 1s digits always end in 0, 2, 4, 6 or 8

724 726 728 730 732 734

734 cans

You can count by 2s from any odd number.

The 1s digits always end in 1 3 5 7 or 9

187 189 191 193 195 197 199 201

201 sheep

Count by 2s from 583 to 607 with your eyes shut.

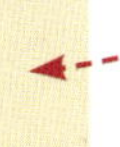

Challenge

Try counting by 20s.
It is like counting by 2s but 10 times more.
There is an extra 0 on each number.

0 20 40 60 80 100 120 140 160 180 200 220

Try counting by 200s.
It is like counting by 2s but 100 times more.
There are two extra 0s on each number.

0 200 400 600 800 1000 1200 1400

COUNT BY 5s TO 1000

You can count by 5s from any number.

The 1s digits always make a pattern

485 490 495 500 505 510

The 1s digit pattern is 5, 0, 5, 0, 5, 0.

510 jellybeans

673 678 683 688 693 698 703 708

The 1s digit pattern is 3, 8, 3, 8, 3, 8.

708 king penguins

Try counting by 50s.

It is like counting by 5s but 10 times more.

There is an extra 0 on each number.

0 50 100 150 200 250 300 350 400

Challenge

Can you count back from 1000 by 50s with your eyes shut?

1000 950 900 850 800 750 700 650 600 550 500 450 400 350 300 250 200 150 100 50 0

COUNT BY 10s TO 1000

You can start from any number when you count by 10s.

740 750 760 770 780 790

The 1s digit pattern is 0 0 0 0 0 0.

790 people

264 274 284 294 304 314 324 334

The 1s digit pattern is 4, 4, 4, 4, 4, 4, 4, 4.

334 lanterns

Try this

Count back by 10s in your head from 482 to 382.

Challenge

Try counting silently by 100s.

It is like counting by 10s but 10 times larger.

There is an extra 0 on each number.

0 100 200 300 400 500 600 700 800 900 1000 1100 ...

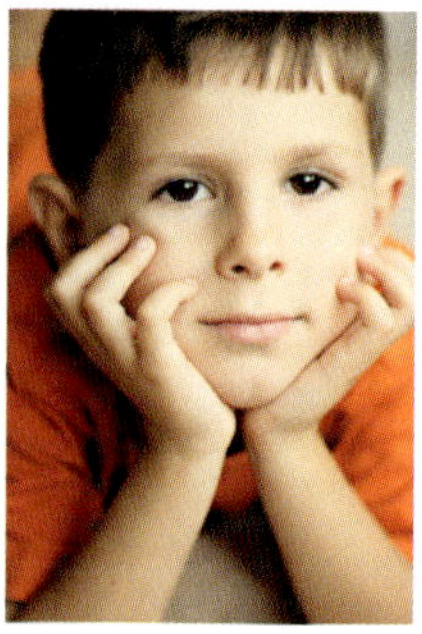

ESTIMATE BY 100

Sometimes you estimate rather than count exactly.

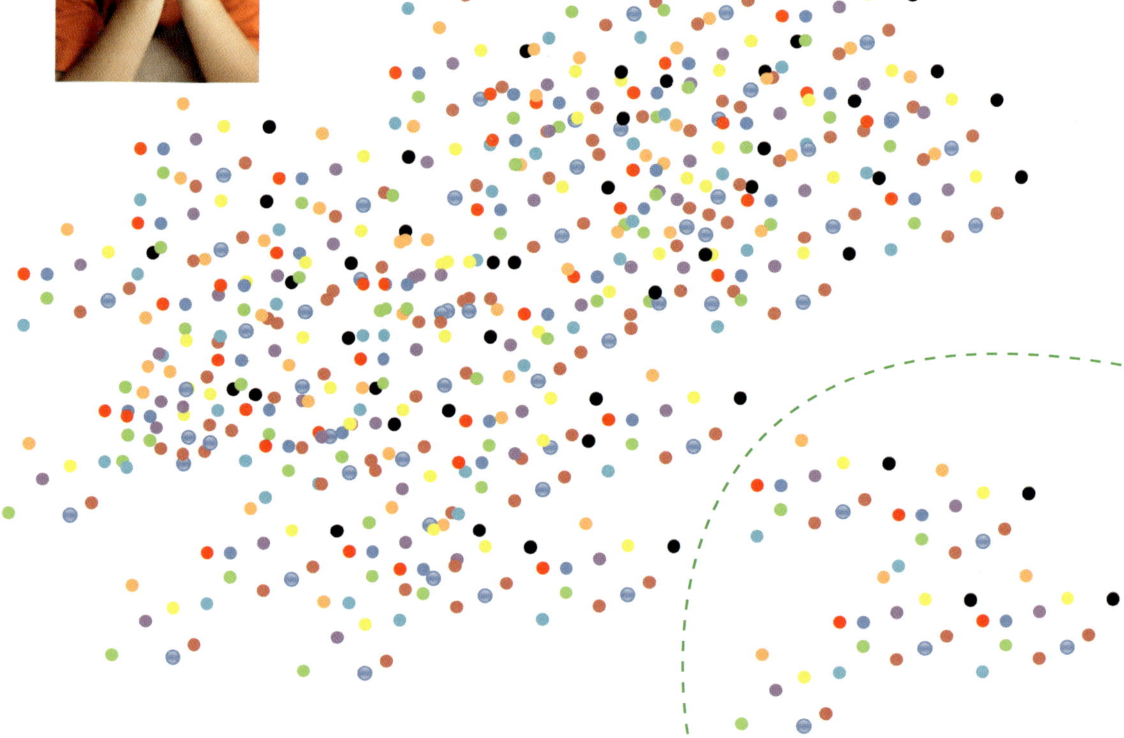

First count a small part of the whole group.
See what 50 looks like.
Imagine more groups this size.
How many groups like this will fit into the whole group?
Count by 50 to find your estimate.

There are 377 coloured dots.
Which estimate was the closest?

PLACE VALUE TO 100

It takes time to count 10 things one by one

GROUPS OF 10

Numbers smaller than a group of 10 are called **ones** or **units**.

0 1 2 3 4 5 6 7 8 9 are all **1-digit** numbers.

It is easier to count one group of 10. **10**

10 is a **2-digit** number.

Use the 0 to 9 digits to write groups of 10.

Use a place value chart to record your numbers.

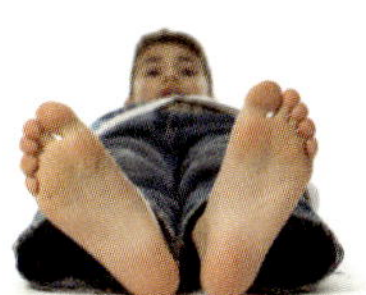

1 ten

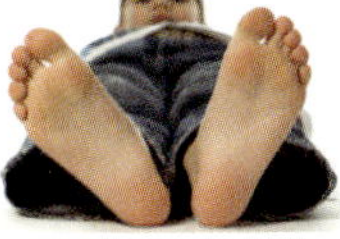

2 tens

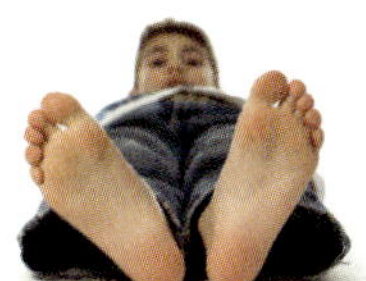

3 tens

10s	1s
3	0

10 ones are the same number as 1 ten

You can count groups of 10.

0 10 20 30 40 50 60 70 80 90 100

10 20 30 40 50

10s	1s
5	0

You can model groups of 10 with beansticks.

10s	1s
5	0

A wombat's burrow can be up to 30 metres long.

GROUPS OF 10 (continued)

You can model groups of 10 with popstick bundles.

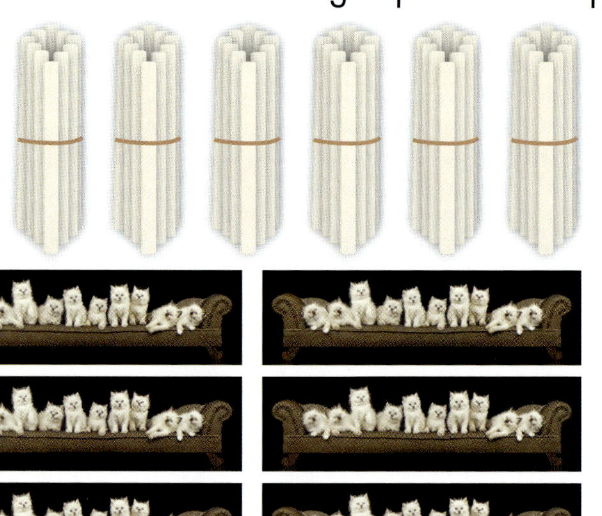

10s	1s
6	0

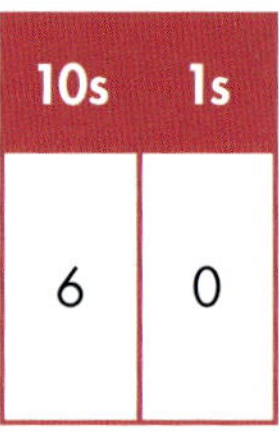

10s	1s
6	0

You can model groups of 10 with blocks.

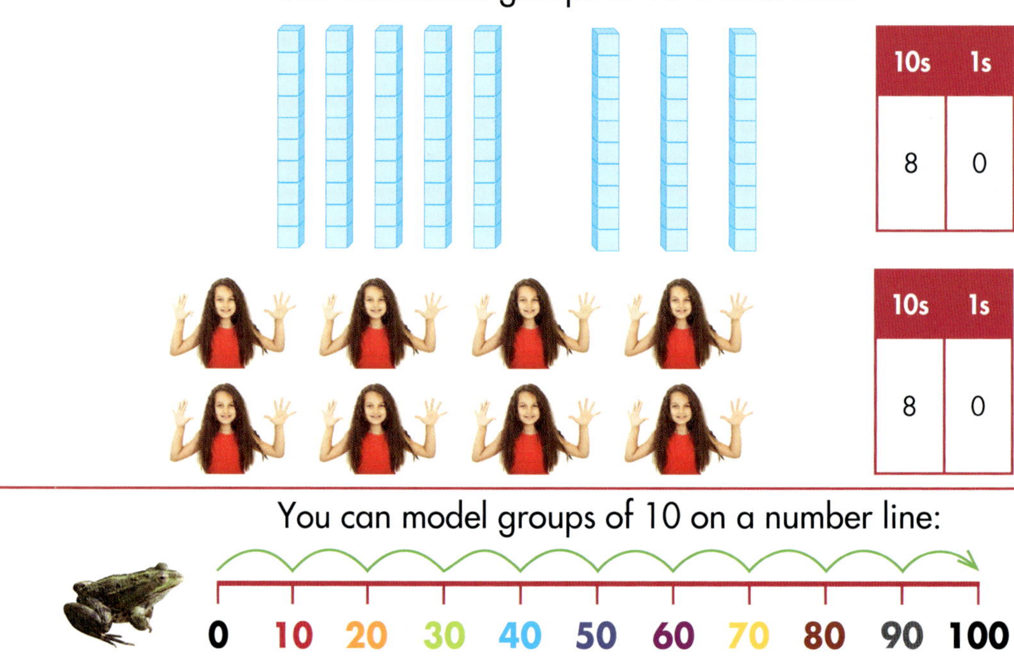

10s	1s
8	0

10s	1s
8	0

You can model groups of 10 on a number line:

0 10 20 30 40 50 60 70 80 90 100

You can sort random groups of 10 from smaller to larger.

30 → 50 → 70 → 90 the numbers get larger

You can sort random groups of 10 from larger to smaller.

80 → 60 → 40 → 20 the numbers get smaller

GROUPS OF 10 SUMMARY

Model	Numeral	Number of 10s	Word
	0	no tens	zero
	10	one ten	ten
	20	two tens	twenty
	30	three tens	thirty
	40	four tens	forty
	50	five tens	fifty
	60	six tens	sixty
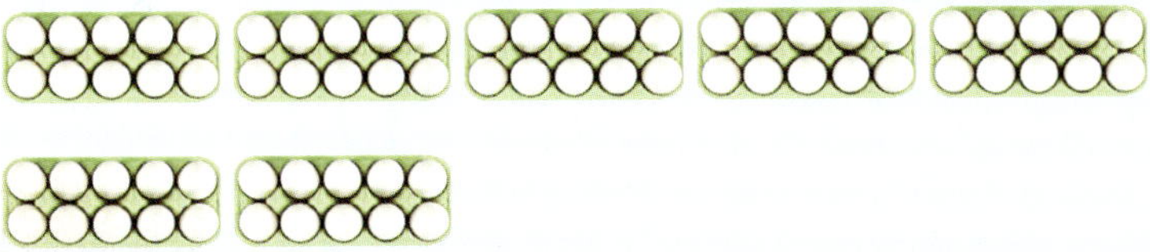	70	seven tens	seventy
	80	eight tens	eighty
	90	nine tens	ninety
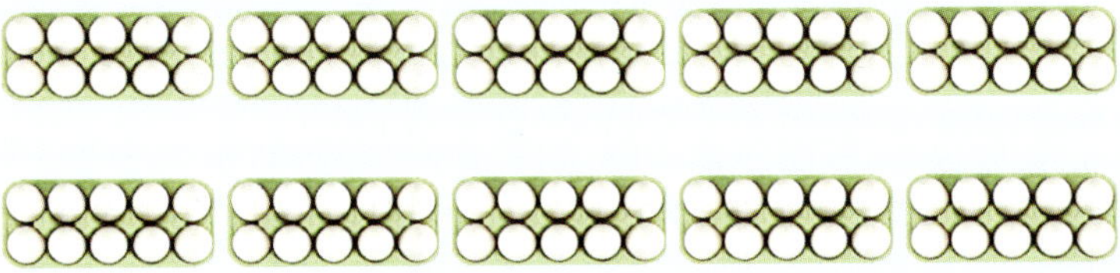	100	ten tens	one hundred

All numbers from 10 to 99 are 2-digit numbers

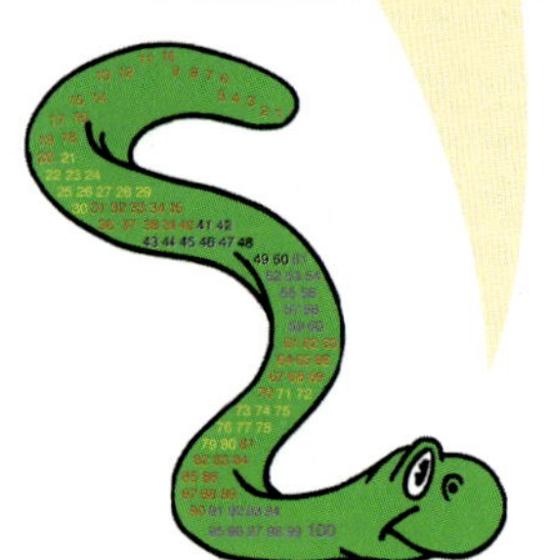

TENS AND ONES

They are made from groups of **tens** and **ones**.

You can model 10s and 1s with beansticks.

10s	1s
2	3

2 is in the 10s place **3** is in the 1s place

2 groups of 10 and 3 extra ones

You can model 10s and 1s with popstick bundles.

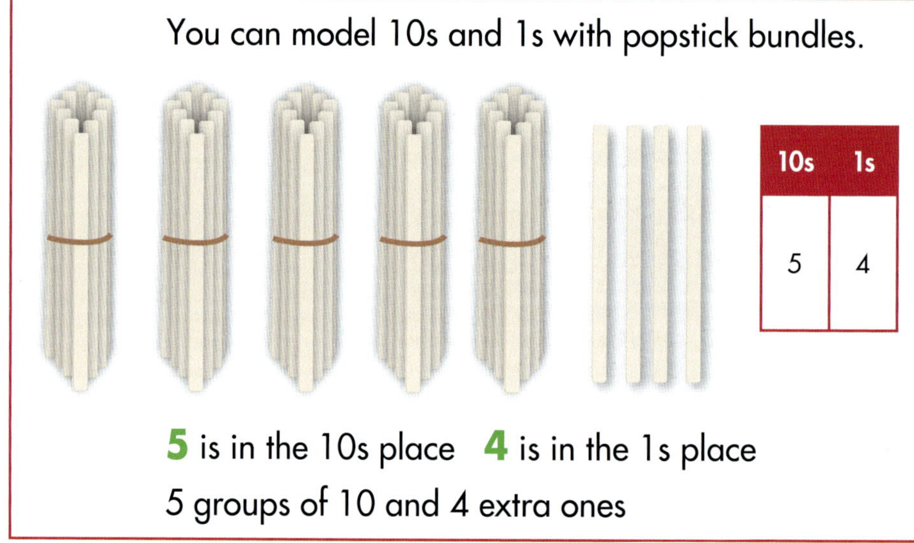

10s	1s
5	4

5 is in the 10s place **4** is in the 1s place

5 groups of 10 and 4 extra ones

You can model 10s and 1s with blocks.

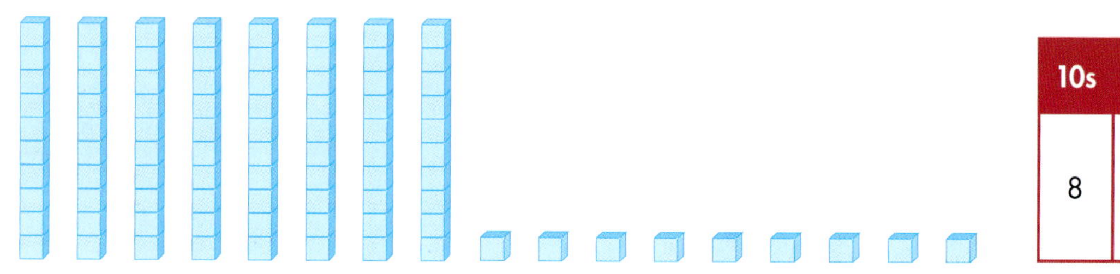

10s	1s
8	9

8 is in the 10s place **9** is in the 1s place

8 groups of 10 and 9 extra ones

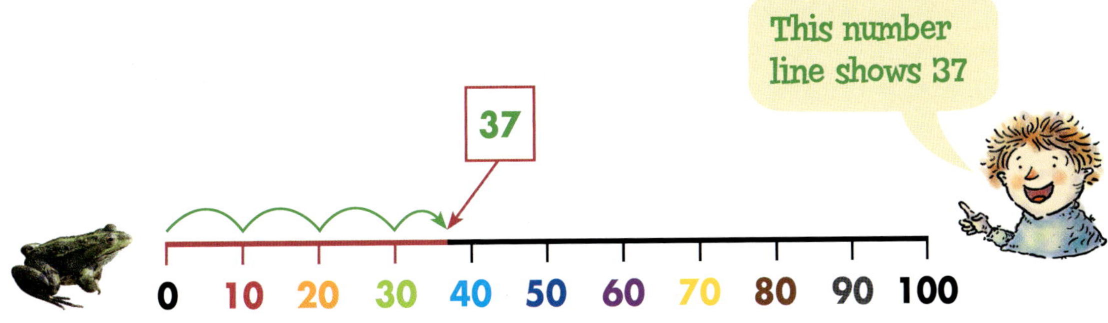

The arrow shows an estimate. It does not have to be exact.

You can say 10s and 1s using different number names.

7 tens and 6 ones is the same amount as 76 ones.

In words this is 'seventy-six'.

10s	1s
7	6
7	6

76 is between 70 and 80.

76 is less than 100.

You can count forwards by 10s from any number.

2 12 22 32 42 52 62 72 82 92

You can count backwards by 10s from any 2-digit number.

93 83 73 63 53 43 33 23 13 3

Try this

How many tenpin bowling pins?

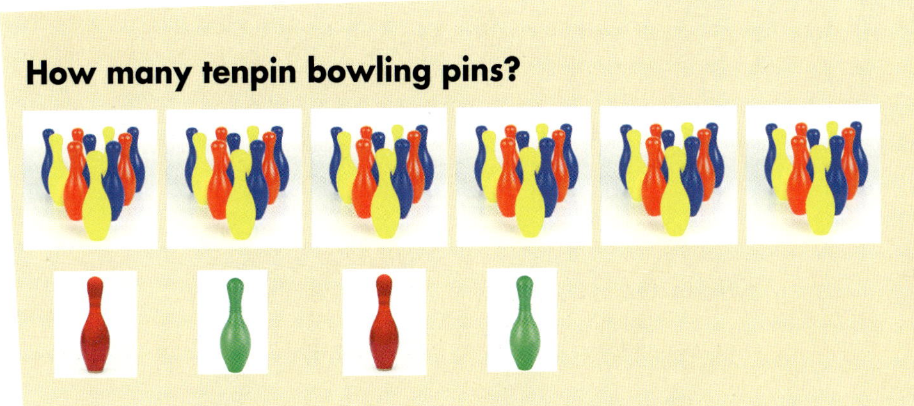

NUMBER & ALGEBRA

ORDERING 2-DIGIT NUMBERS

A 1-digit number is always smaller than a 2-digit number

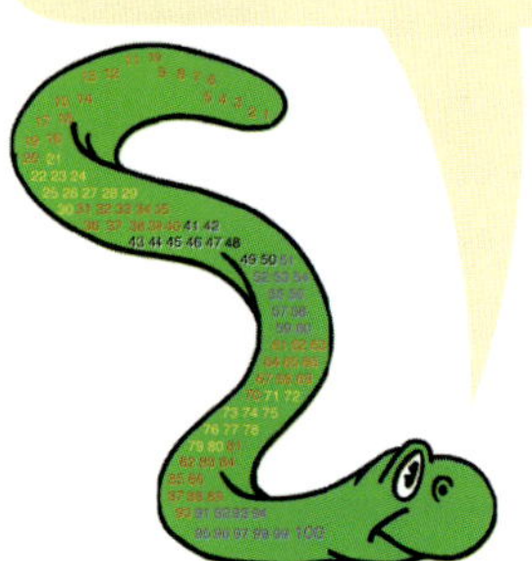

6 is smaller than **29**

6 **29**

You can sort random 2-digit numbers into order counting forwards.

Look at the number in the 10s place.

The number in the 10s place gets larger

27 40 85 92

You can sort random 2-digit numbers in order counting backwards.

Look at the number in the 10s place.

93 77 59 34

The number in the 10s place gets smaller

If two numbers have the same 10s digit, look at the 1s digit.

The number with the smaller 1s digit is the smaller number.

67 and 69

7 is smaller than **9** so 67 is smaller than 69

Sort these numbers into order counting forwards.

81, 31, 78, 56 _____ _____ _____ _____

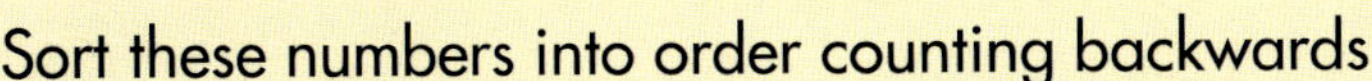

Sort these numbers into order counting backwards.

49, 95, 36, 26 _____ _____ _____ _____

ROUND TO THE NEAREST 10

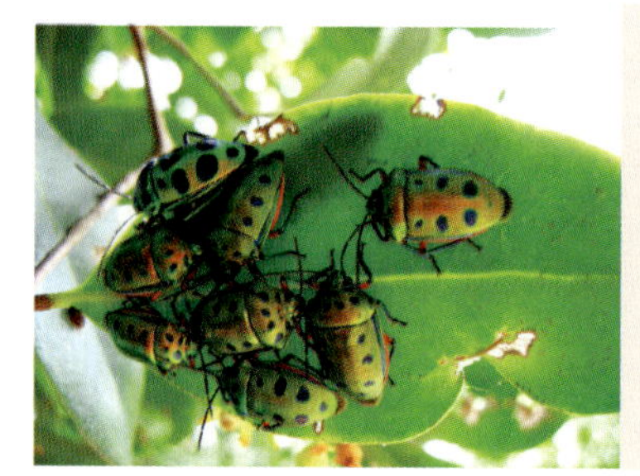

Counting doesn't always need to be exact.

How many beetles are here? There are exactly 8.

But if you round to the nearest 10 you can say 'about 10'.

This is useful when you estimate.

When you estimate, you want to make the 1s digit 0.

Which group of 10 is closest?

To decide you look at the ones digit.

5 6 7 8 9 are closer to 10 so round **up**.

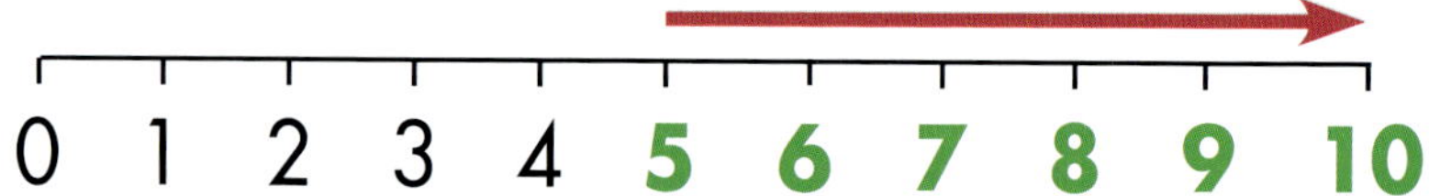

If the ones digit is **5 or more**, round it **up** to the next multiple of 10 and write 0 ones.

27 is closer to 30

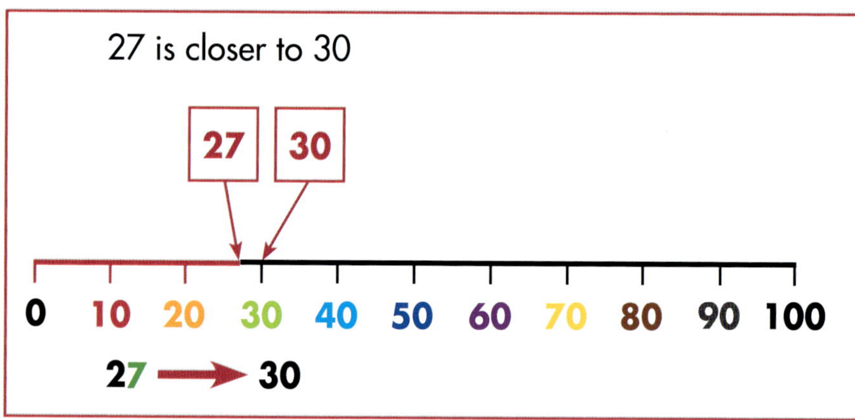

There are **about 30** oranges in this box.

55 → **60** 55 is closer to 60

89 → **90** 89 is closer to 90

0 1 2 3 4 are closer to 0 so round **down**.

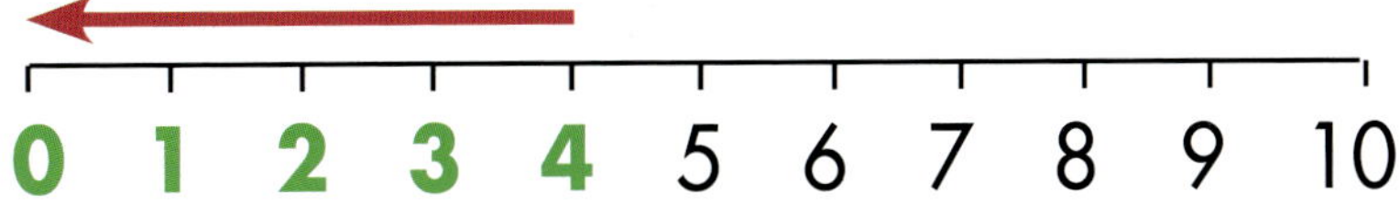

If the ones digit is **less than 5**, round it **down** to 0.
The 10s digit stays the same.

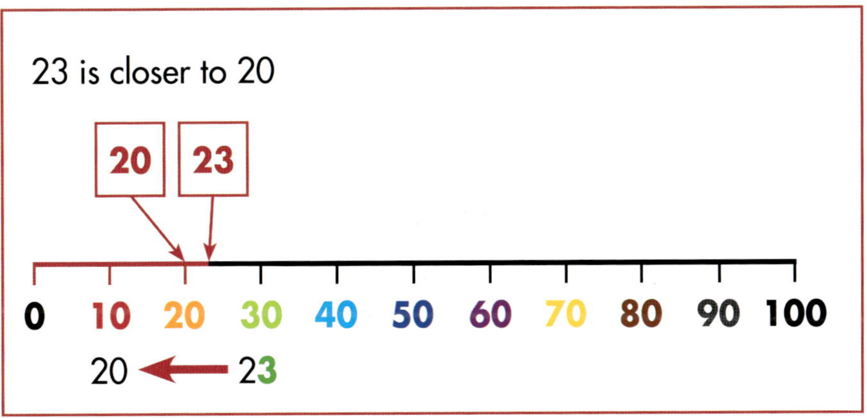

40 ← **44** 44 is closer to 40.

This chameleon is **about 40 cm** long.

70 ← **72** 72 is closer to 70

90 ← **93** 93 is closer to 90

Try this

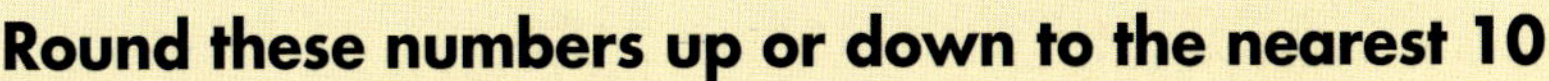

Round these numbers up or down to the nearest 10

41 ______ **68** ______ **99** ______ **53** ______

I love hundreds!

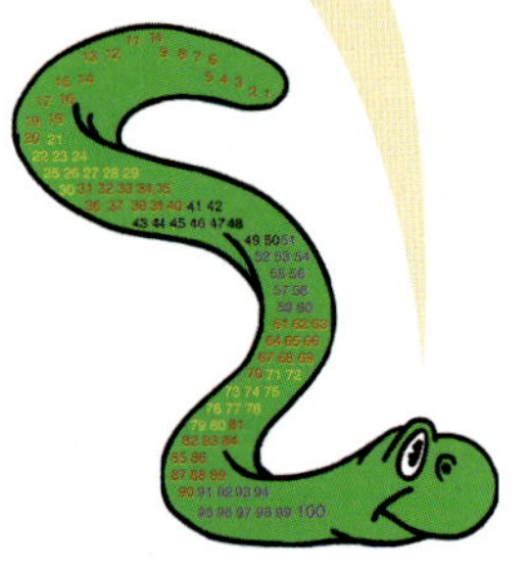

PLACE VALUE TO 1000

HUNDREDS

Every time you collect ten groups of 10, you have a new unit called a **hundred**.

10 tens are the same amount as one 100

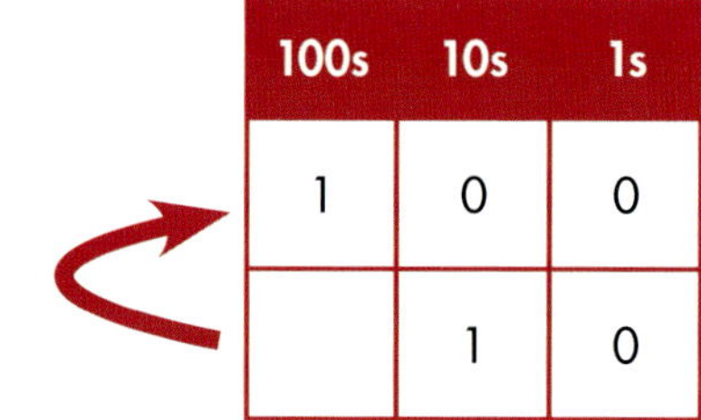

100s	10s	1s
1	0	0
	1	0

100 is a **3-digit** number

100 is ten times larger than 10

You can model 100s with blocks.

10 'longs'

10 tens

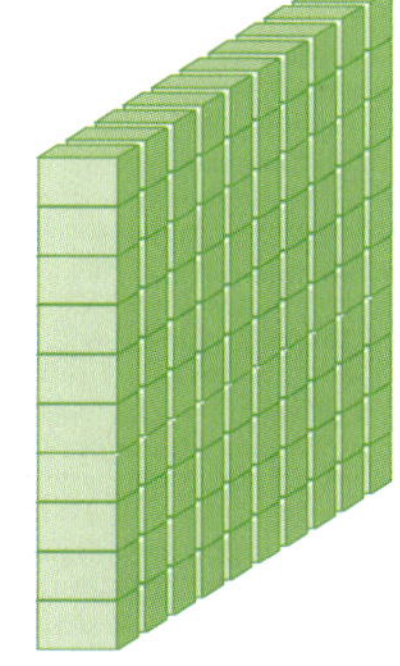

1 'flat'

1 hundred

10 'longs' are the same amount as 1 'flat'. You can pick up 1 'flat' more easily than you can pick up 10 'longs'. You can swap 10 'longs' for 1 'flat'.

You can model 100s with an abacus.

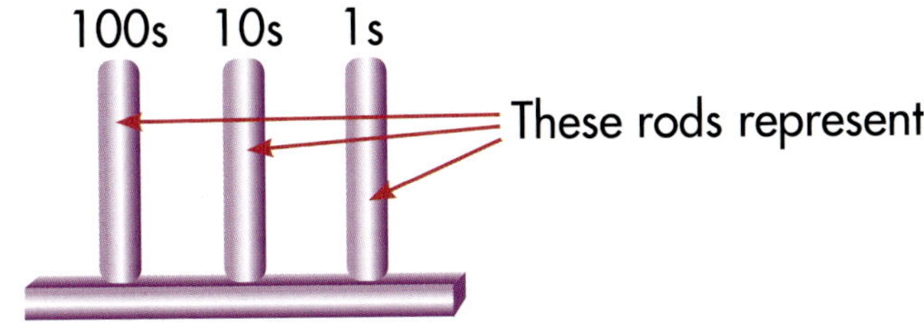

An abacus uses beads to model a number.

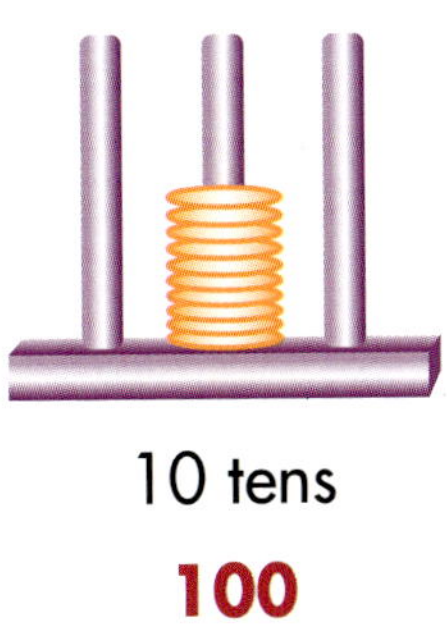

10 tens

100

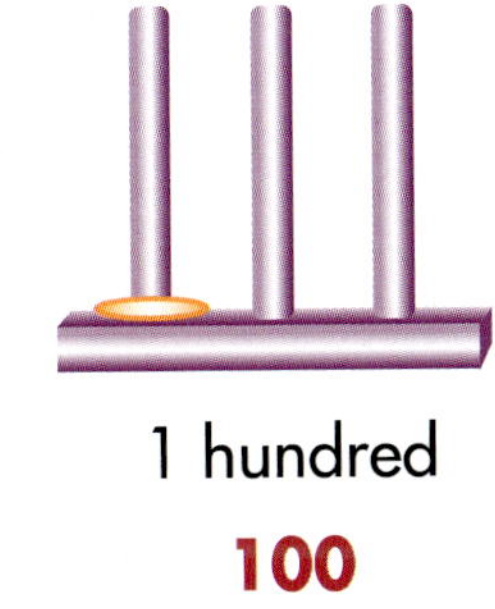

1 hundred

100

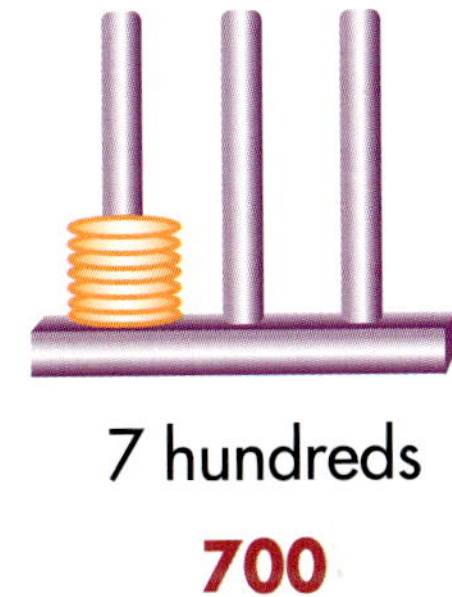

7 hundreds

700

You can count forwards by hundreds from 0.

0 100 200 300 400 500
600 700 800 900

100 cents = $1

You can count backwards by hundreds from 900.

900 800 700 600 500 400
300 200 100 0

One hundred $1 coins has the same value as a $100 note.

All numbers from 100 to 999 are 3-digit numbers

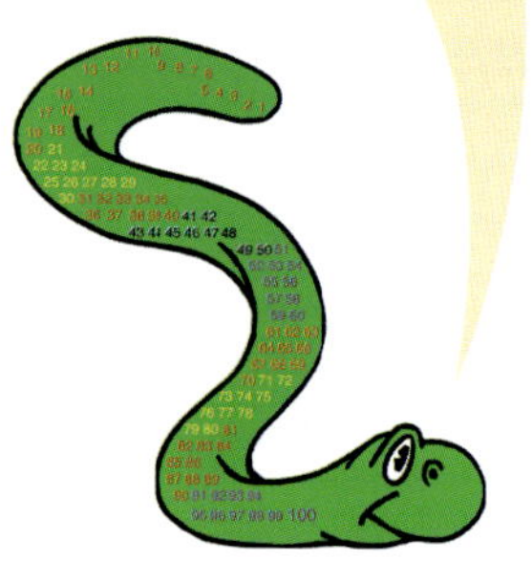

HUNDREDS, TENS AND ONES

All 3-digit numbers are made from groups of **hundreds**, **tens** and **ones**.

You can model 100s, 10s and 1s with blocks.

3	8	2
hundreds	tens	ones

3	8	2

3 is in the 100s place, 8 is in the 10s place, 2 is in the 1s place.

You can model 100s, 10s and 1s with an abacus.

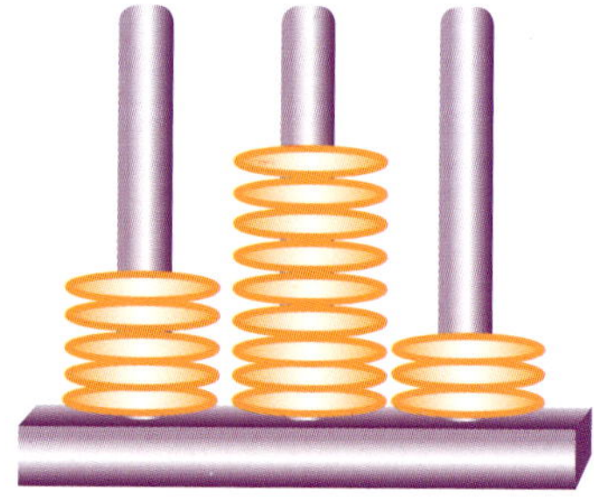

5 hundreds 9 tens 3 ones

5 is in the 100s place

5	9	3

Chinese people have used an abacus to model numbers for over 800 years.

You place beads on each rod of the abacus to model the hundreds, tens and ones of your number.

You can draw 100s, 10s and 1s as jumps along a number line.

0 100 200 300 400 500 600 700 800 900 1000

The arrow shows an estimate.

It does not have to be exact.

You can say 100s, 10s and 1s using different number names.

Hundreds	Tens	Ones	
7	3	4	7 hundreds, 3 tens and 4 ones
7	3	4	73 tens and 4 ones
7	3	4	734 ones

In words this is 'seven hundred and thirty-four'.

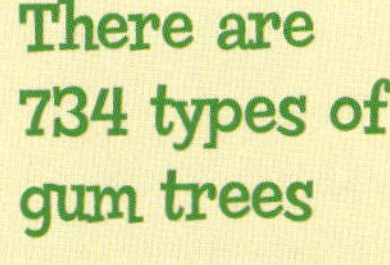

You can count forwards by 100s from any number.

271 371 471 571 671 771 871 971

You can count backwards by 100s from any number.

856 756 656 556 456 356 256 156 56

Count forward by 100s starting from 162.

Count backwards by 100s starting from 983.

Try this

ORDERING 3-DIGIT NUMBERS

A 1-digit number is always smaller than a 2-digit number.

A 2-digit number is always smaller than a 3-digit number.

You can sort 3-digit numbers in order counting forwards.

Look at the number in the **100s** place.

349 481 503 722

One person uses about 249 litres of water each day.

The number in the 100s place gets larger as you sort forwards.

You can sort 3-digit numbers in order counting backwards.

Look at the number in the **100s** place.

802 694 372 249

The number in the 100s place gets smaller as you sort backwards.

If two numbers have the same 100s digit, look at the 10s digit

The number with the smaller **10s** digit is the smaller number.

457 and **482**

5 is smaller than 8 so 457 is smaller than 482.

My dog walked 643 steps

If two numbers have the same 100s digit and the same 10s digit, look at the **1s** digit.

The number with the smaller **1s** digit is the smaller number.

643 and **646**

3 is smaller than 6 so 643 is smaller than 646.

REARRANGING 3-DIGIT NUMBERS

Create rules to rearrange and make different 3-digit numbers using three-digit cards.

			the largest number
			a number between 800 and 850
			the number closest to 500
			the smallest number

Try this

Grandpa's operation cost $790

Rearrange these 3 digits to make:

the number closest to 900 ______

the smallest number ______

a number between 600 and 800 ______

ROUND TO THE NEAREST 100

Counting doesn't always need to be exact. You can round 3-digit numbers to the nearest 100. It is easier to count and calculate with 100s.

Here are exactly 381 dog munchies.

But if you round to the nearest 100 you can say 'about 400'.

This is useful when you estimate.

To decide you look at the 10s and 1s digits

You want to put 0 in the 10s place.

You want to put 0 in the 1s place.

Which group of 100 is the closest?

50 or above? Round up

If the number is **50 or more**, round it **up** to the next 100 and write 0 in the 10s and 1s place.

462 → **500** 462 is closer to 500

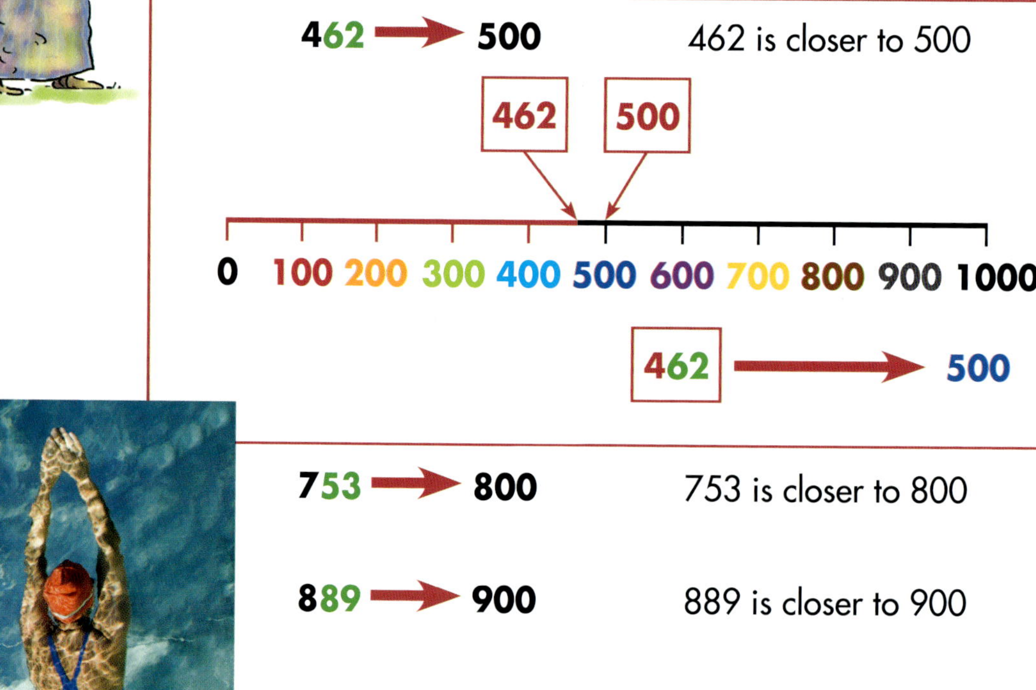

753 → **800** 753 is closer to 800

889 → **900** 889 is closer to 900

Mum swims about 500 metres every day

Less than 50? Round down

If the number is **less than 50**, round it **down** to 0 in the 10s and 1s place.

The hundreds digit stays the same.

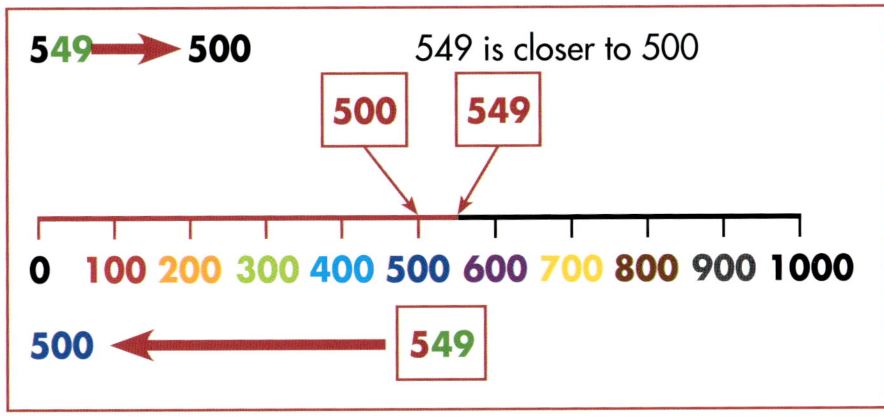

There are about 500 books in our library.

600 ← 627 627 is closer to 600

700 ← 705 705 is closer to 700

Try this

Round these numbers up or down to the nearest 100.

Explain your reasons to a friend.

519 ______

762 ______

486 ______

255 ______

THOUSANDS

10 hundreds make 1000

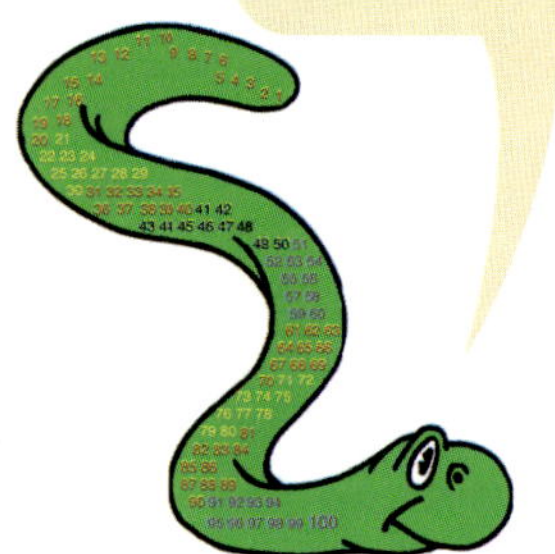

Every time you collect ten groups of 100, you get a new unit called a **thousand**.

You can model 1000s, 100s, 10s and 1s with blocks.

1 block

1 thousand

1000

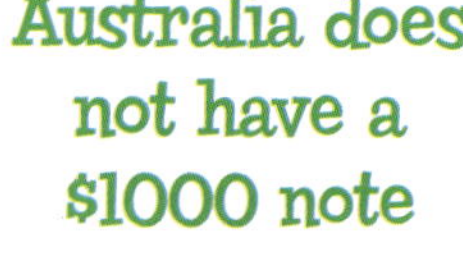

1000s	100s	10s	1s

1000 is a **4-digit** number

1000 is ten times larger than 100

10 'flats' are the same amount as 1 'block'. You can pick up 1 'block' more easily than you can pick up 10 'flats'. You can swap 10 'flats' for 1 'block'.

Some special groups of 1000

1000 cents = $10

1000 grams = 1 kilogram

1000 millilitres = 1 litre

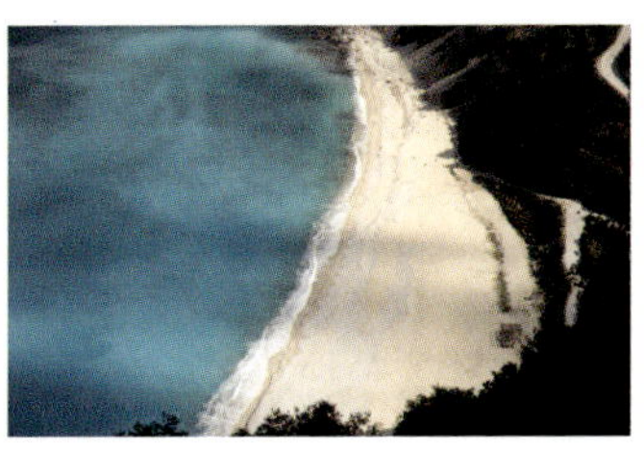

1000 metres = 1 kilometre

1 kilometre long beach at Myrtos, Greece

You can record 1000s on an abacus.

The 4th rod represents groups of 1000.

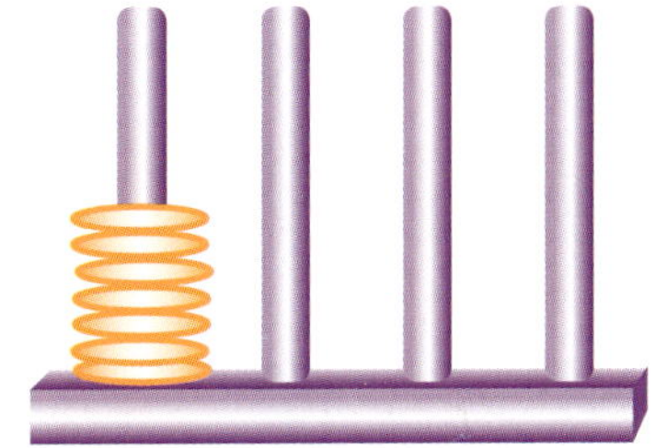

You can record 1000s on a number line.

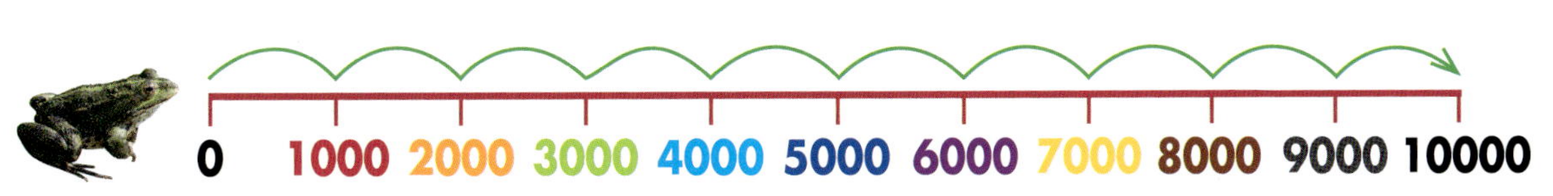

You can count forwards by thousands from 0.

0 1000 2000 3000 4000 5000 6000 7000 8000 9000

You can count backwards by thousands from 9000.

9000 8000 7000 6000 5000 4000 3000 2000 1000 0

ADD AND SUBTRACT TO 10

ADD TO 10

To add, you join 2 or more groups together to make a larger group. You increase the size.

Lucy has 3 pups and 2 pups

3 and **2** more makes **5**

It doesn't matter which group you add first.

2 and **3** also makes **5**

Lucy has 5 pups altogether.

You can use a 10s frame to help you add.

It helps you learn your facts to 10.

5 and **3** makes **8**

If you add 2 more counters, you'll have 10 altogether.

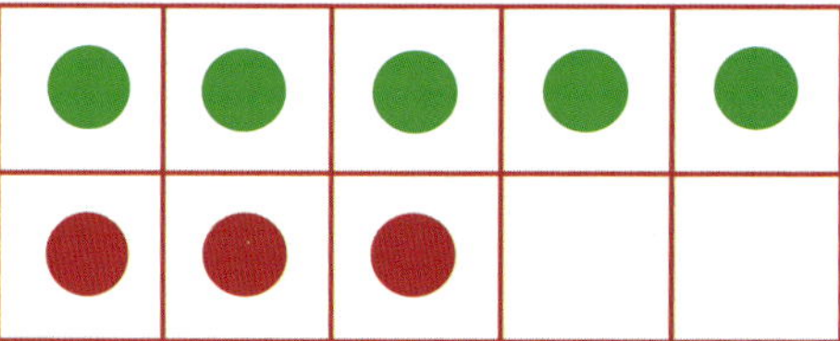

You can use a number line to record your addition story.

You jump forwards.

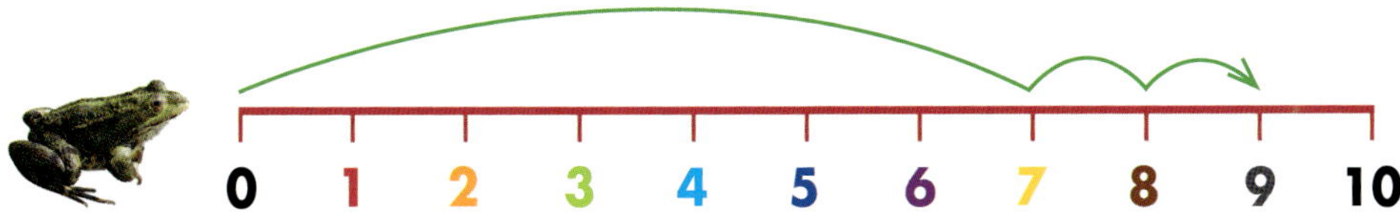

7 and **2** more makes **9**

You can jump once or many times.

If you add 1 more jump, there will be 10.

If you add **0**, you don't add anything at all.

The number in the group stays the same.

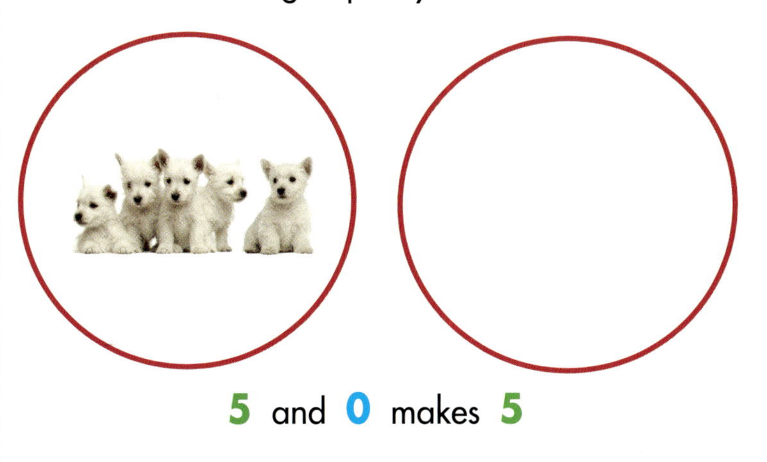

5 and **0** makes **5**

4 and 0 makes … ______

6 and 0 makes … ______

3 and 0 more makes … ______

8 and 0 more makes … ______

COUNT ON

Max has 6 fish.

6

Gran gives him 4 more fish.

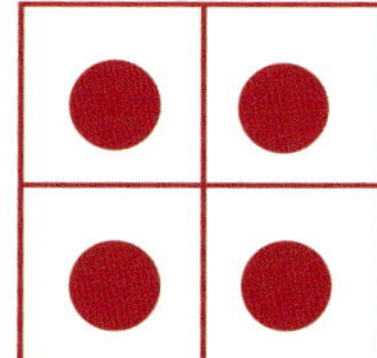

Start at the larger number and count on 4 more.

6 and 4 more makes 10.

Max now has 

fish

Sometimes you don't know how many are in each group. It's like a puzzle.

Ella has 4 scoops in 1 cone and some scoops in another cone.

Altogether Ella has 7 scoops.

4 and ☐ makes 7

How many scoops are on the other cone?

Count on until you get 7.

4 5 6 7

You count on 3 more.

So Ella has 3 scoops in the other cone.

4 and 

makes 7

ADDITION WORDS AND SYMBOLS

Here are different ways to talk about adding:

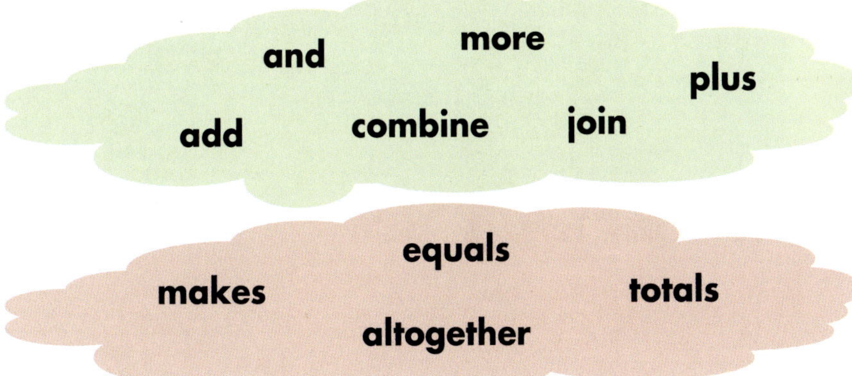

Mathematicians say 'plus' instead of 'and'

Mathematicians say 'equals' or 'total' instead of 'makes'

5 plus **1** equals **6** 6 faces

Mathematicians write the symbol + instead of 'plus'

Mathematicians write the symbol = instead of 'equals' or 'totals'

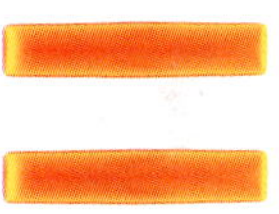

4 + **5** = **9**

9 boomerangs

4 + 5 = 9 is a number sentence.

It tells a story about addition.

You can also write it this way.

9 = 4 + 5

Make up your own word story about 2 + 7 = 9

ADDITION FACTS ABOUT 10

There are 11 ways to combine 2 groups to make 10.

You are a Maths Star if you can remember these combinations

ADDING MORE THAN 2 GROUPS

You can add more than 2 groups to make 10 or more.

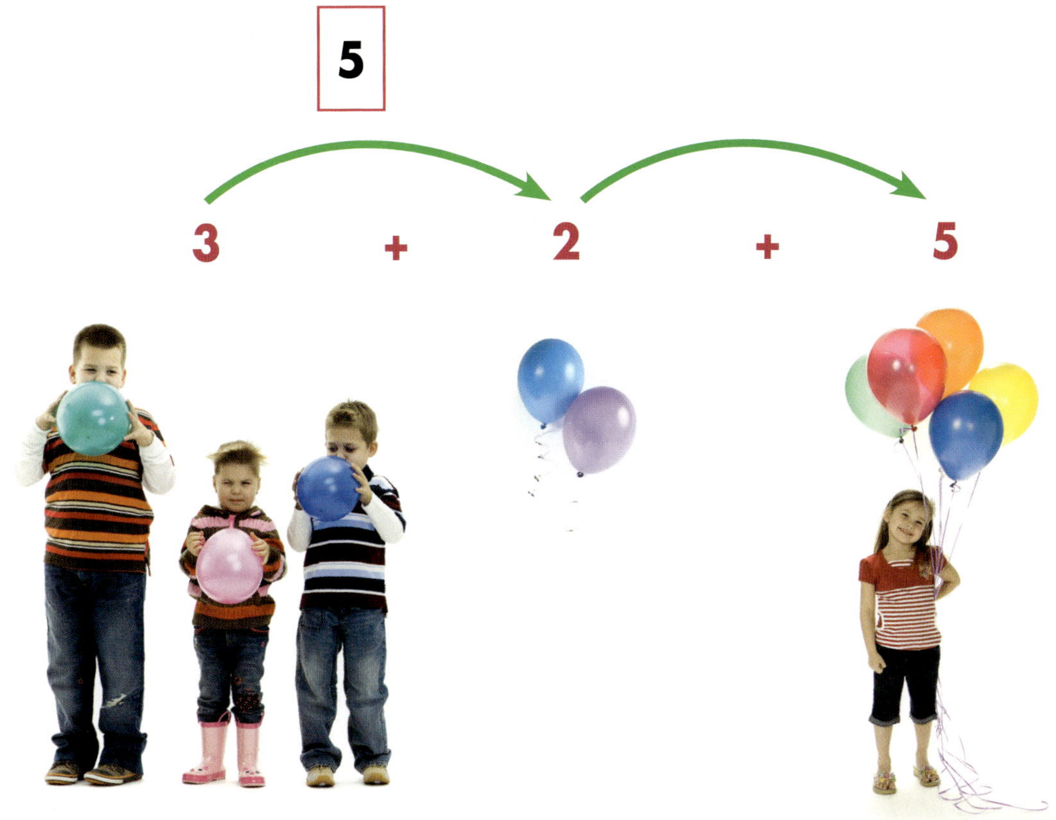

There are now **10** balloons.

Josh throws these dice.

1 + 4 + 5 = 10

Josh scores 10.

You can add more numbers like this forever.

Here are a few more combinations that equal 10.

3 + 6 + 1 = 10

5 + 2 + 2 + 1 = 10

4 + 3 + 2 + 1 + 0 = 10

TAKE AWAY FROM 10

To take away, you remove things from a larger group.

5 take away 2 leaves 3

You cross out 2 puppets.

The number in the group gets smaller.

You decrease the size.

If you take away 3 more puppets, there will be 0 left.

10 take away 6 leaves 4

If you take away 4 more leaves, there'll be none left.

You can use a 10s frame to help you take away.

It helps you learn your facts up to 10.

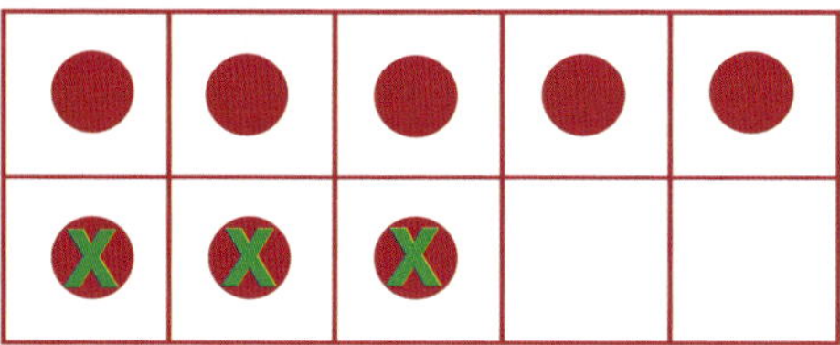

8 take away 3 leaves 5

If I take away 5 more there will be none left.

You can use a number line to record your take away story.

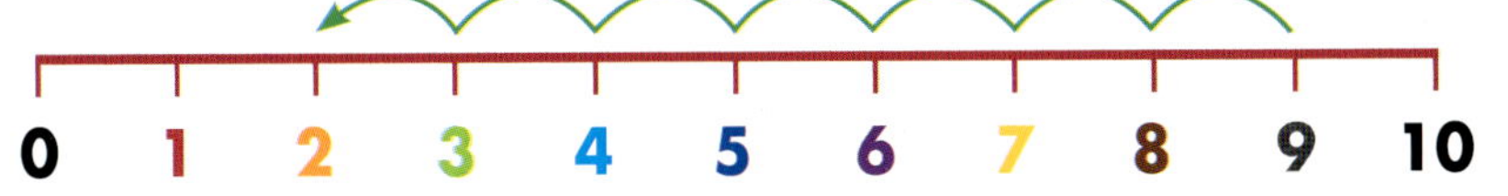

9 take away 7 leaves 2

If I take away 2 more it would leave 0.

Try this

1. Show 9 take away 4 on this 10s frame.

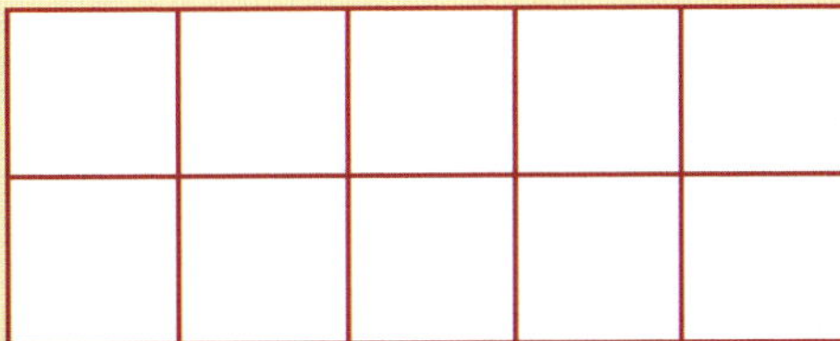

2. Draw jumps to show 10 take away 5.

If you take away **0** you still have the same amount you started with.

Jack has 7 books to read. He hasn't read any of them. He still has 7 books to read.

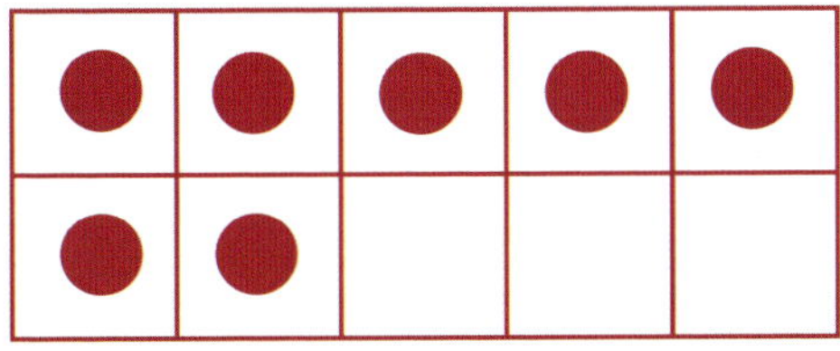

COUNT BACK

Nada has 6 puppets.

She gives 4 away.

Start at the larger number and count back 4 more.

 **5 4 3 2**

Nada has 2 puppets left.

You call this 'counting back'

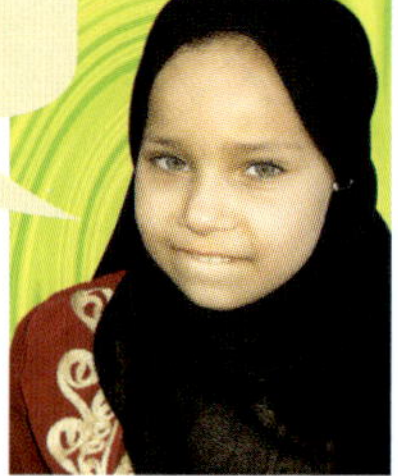

Sometimes you don't know how many are in each group.

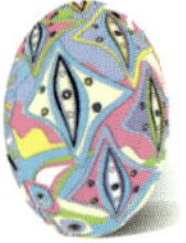

It's like a puzzle.

Chen has 10 eggs.
He gives some to his sister.
He now has 7 eggs left.

10 take away  leaves 7

How many eggs does he give his sister?

Count back until you get 7.

 **9 8 7**

You count back 3.

Chen gives his sister 3 eggs.

10 take away 3 leaves 7

8 take away  leaves 3

9 take away  leaves 2

SUBTRACTION WORDS AND SYMBOLS

Mathematicians say 'minus' or 'subtract' instead of 'take away'.

Mathematicians say 'equals' instead of 'leaves'.

9 minus **3** equals **6**

Mathematicians write the symbol – instead of 'take away'.

Mathematicians write the symbol = instead of 'equals'.

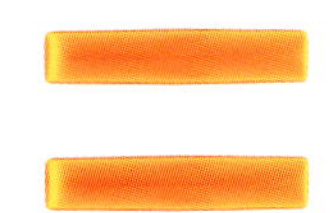

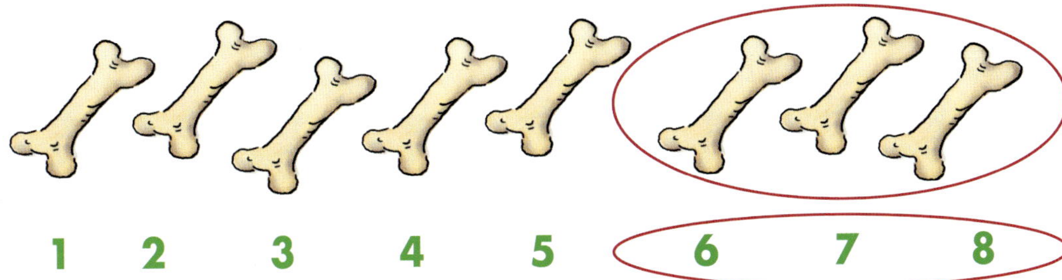

8 – 3 = 5

8 – **3** = **5** is a number sentence.

It tells a story about subtraction.

You can also write it this way.

5 = **8** – **3**

NUMBER FACT FAMILIES

The funny thing is you can tell the same story in many different ways. You can make a fact family about each number sentence.

You know 4 addition facts.

3 + 5 = 8
5 + 3 = 8
8 = 5 + 3
8 = 3 + 5

You know 4 subtraction facts.

8 – 3 = 5
8 – 5 = 3
5 = 8 – 3
3 = 8 – 5

You know 8 facts altogether

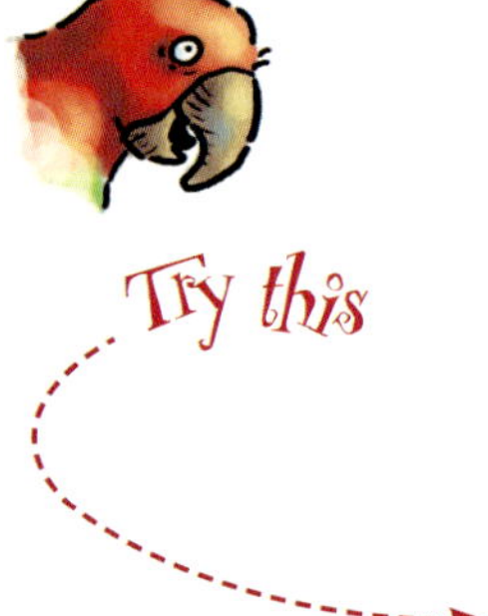

Try this

Write 2 adding facts for 4 = 7 – 3

Write 2 take away facts for 1 + 9 = 10

Challenge

Look at page 40.

You are a Maths Star when you can remember all these combinations as subtraction facts too.

ADD AND SUBTRACT TO 20

When you add, you do addition.

9 + 6 = 15

But be careful

When you take away, you do subtraction.

15 – 6 = 9

Subtraction is the reverse of addition.

Every time you know an addition fact you also know a subtraction fact.

When you add, it doesn't matter which number comes first.

9 + 6 is the same number as 6 + 9.

When you subtract, it **does** matter which number comes first.

15 – 6 is **not** the same number as 6 – 15.

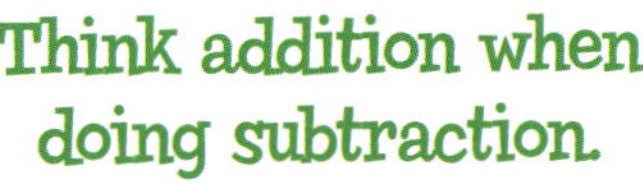

17 – 8 = □

is the same as saying

8 + □ = 17

+	add
=	is the same number as
sum	the total or answer to an addition problem

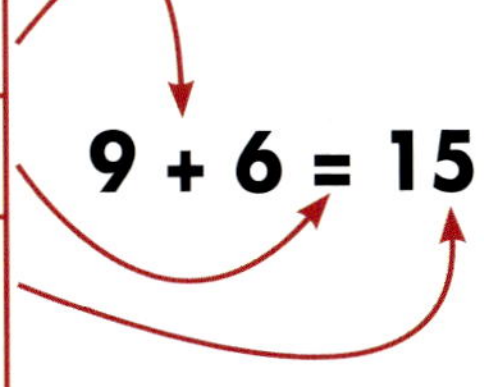

–	subtract
=	is the same number as
difference	the answer to a subtraction problem

6 = 15 – 9

MENTAL STRATEGIES

You can put 20 objects into many different size groups.

If you can add and subtract numbers in your head, you don't need to use your fingers.

Try these strategies

It's easy to add another number to 10.

Look at the pattern.

Each number has a 1 in the 10s place.

The 1s digit stays the same.

10 +	0	1	2	3	4	5	6	7	8	9
	10	11	12	13	14	15	16	17	18	19

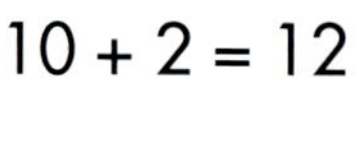

10 + 2 = 12

10 + 7 = 17

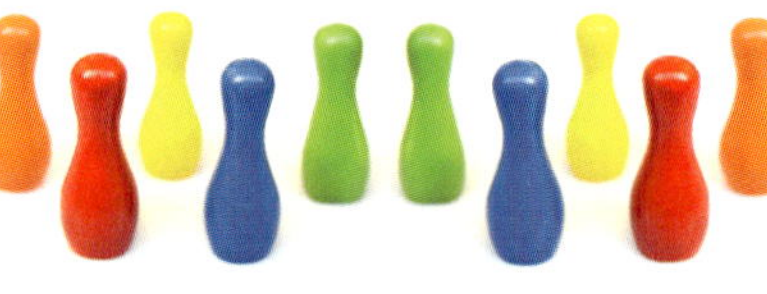

4 + 10 = 14

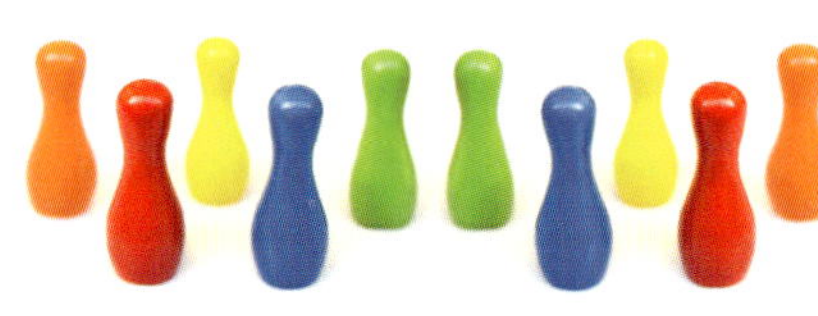

But be careful

Your answer is the same as when you add a number on to 10.

0 + 10 is the same number as 10 + 0.

But be careful.

It is not true in subtraction.

10 – 0 is **not** the same number as 0 – 10.

Build to 10

Imagine a 10s frame in your head.

6 plus 9

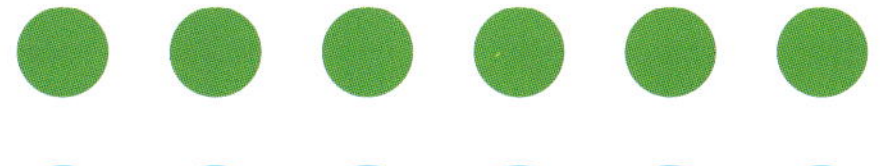

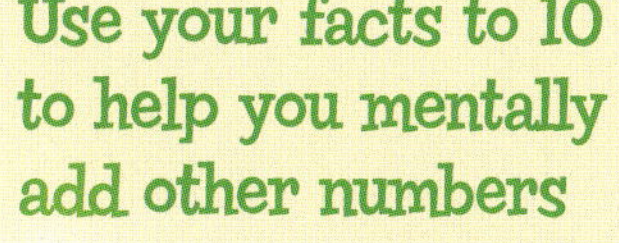

You know 6 + 4 = 10.

That still leaves 5.

Just add this 5 to 10 to get 15.

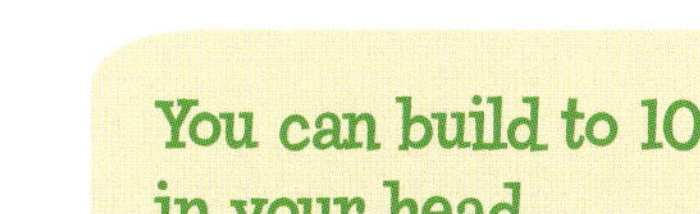

Ali has 8 thongs.

Raja has 4 thongs.

How many thongs altogether?

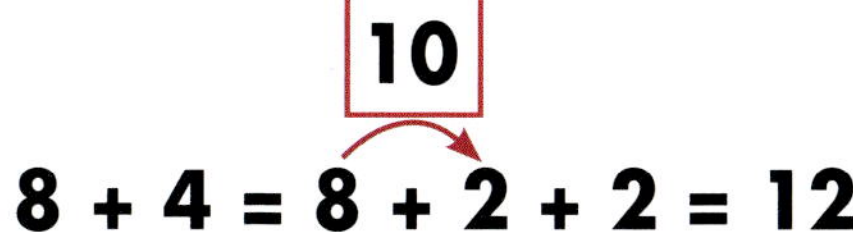

NUMBER & ALGEBRA

Doubles

2 ×

double

2 groups of

2 times

twice as many

Double	0	1	2	3	4	5	6	7	8	9	10
	0	2	4	6	8	10	12	14	16	18	20

Look at the pattern.

Whole number doubles are even numbers.

If you can't remember them, try the 'build to 10' strategy.

Near doubles

Use your doubles facts to 10 + 10 or 20 – 10 to work out near doubles. These are number pairs with a difference of 1, like 5 + 6 or 9 + 10.

Double the smaller number then add 1.

10

6 + 5 = 5 + 5 + 1 = 11

Or double the larger number then subtract 1.

18

8 + 9 = 9 + 9 – 1

There are 17 leaves altogether.

FACTS TO 20 CHART

+/-	0	1	2	3	4	5	6	7	8	9	10
0	0	1	2	3	4	5	6	7	8	9	10

+/-	0	1	2	3	4	5	6	7	8	9	10
1	1	2	3	4	5	6	7	8	9	10	11

+/-	0	1	2	3	4	5	6	7	8	9	10
2	2	3	4	5	6	7	8	9	10	11	12

+/-	0	1	2	3	4	5	6	7	8	9	10
3	3	4	5	6	7	8	9	10	11	12	13

+/-	0	1	2	3	4	5	6	7	8	9	10
4	4	5	6	7	8	9	10	11	12	13	14

+/-	0	1	2	3	4	5	6	7	8	9	10
5	5	6	7	8	9	10	11	12	13	14	15

+/-	0	1	2	3	4	5	6	7	8	9	10
6	6	7	8	9	10	11	12	13	14	15	16

+/-	0	1	2	3	4	5	6	7	8	9	10
7	7	8	9	10	11	12	13	14	15	16	17

+/-	0	1	2	3	4	5	6	7	8	9	10
8	8	9	10	11	12	13	14	15	16	17	18

+/-	0	1	2	3	4	5	6	7	8	9	10
9	9	10	11	12	13	14	15	16	17	18	19

+/-	0	1	2	3	4	5	6	7	8	9	10
10	10	11	12	13	14	15	16	17	18	19	20

You are a Maths Star when you remember these facts

+	7
6	13

To find an **addition** fact, look at the white space where a yellow number meets a blue number.

6 + 7 = 13 and 7 + 6 = 13

-	8
9	17

To find a **subtraction** fact, start at a white space number and read back and up to the matching yellow and blue numbers.

17 − 9 = 8 and 17 − 8 = 9

Challenge

How many different addition and subtraction number sentences can you write about 14?

MULTIPLY AND DIVIDE

It only works if you add the same number each time

HOW TO MULTIPLY

Multiplication is a fast way to add.

It is repeated addition.

Molly saw 5 lizards in her garden.

How many legs is that?

Molly draws this picture to work it out.

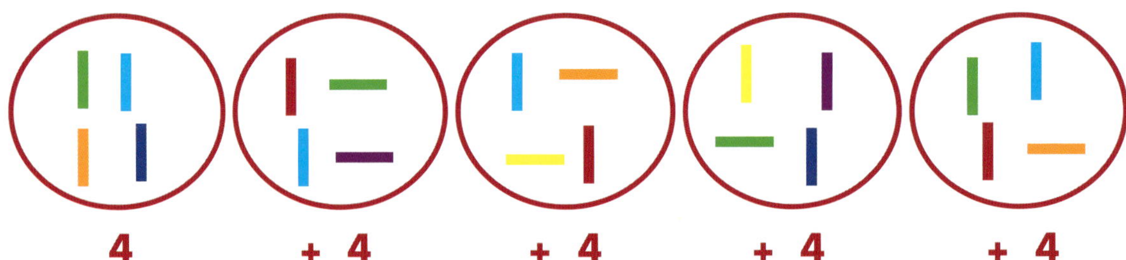

4 + 4 + 4 + 4 + 4

Molly counts by 4s.

4 → 8 → 12 → 16 → 20

5 groups of 4 makes 20

Molly saw 20 lizard legs.

Wayan sees 6 nests.
Each nest has 3 eggs.
How many eggs is that?
6 groups of 3 eggs

Wayan imagines the eggs in **rows**.
There are 3 eggs in every row.
The rows don't have to be equal lengths.
6 rows of 3 eggs makes 18 eggs

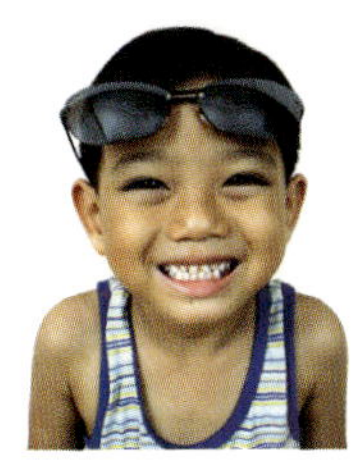

Wayan imagines the eggs in **columns**.
There are 3 eggs in every column.
The columns don't have to be equal lengths.
6 columns of 3 eggs is equal to 18 eggs
Wayan counts by 3s to help him multiply.

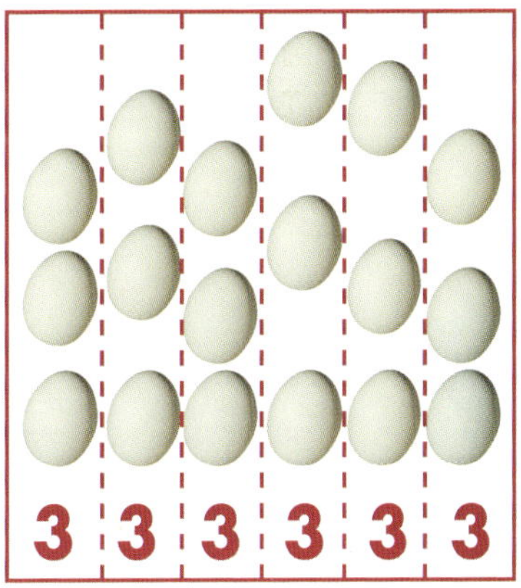

3 + 3 + 3 + 3 + 3 + 3

3 → 6 → 9 → 12 → 15 → 18

Wayan saw 18 eggs altogether.

MULTIPLICATION WORDS AND SYMBOLS

Mathematicians write the symbol **×**
instead of 'groups of', 'rows of' or 'columns of'.
Mathematicians write the symbol **=**
instead of 'makes', 'equals' or 'is equal to'.

4 × 3 = 12 is a **number sentence**.
It tells a story about multiplication.

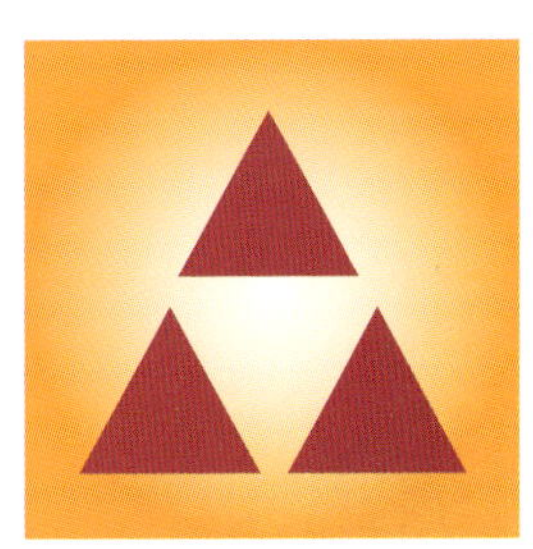

RECORD ON A NUMBER LINE

A number line helps you record a multiplication story.

Ava buys 4 lunch boxes.

Each box has 3 rolls inside.

How many rolls altogether?

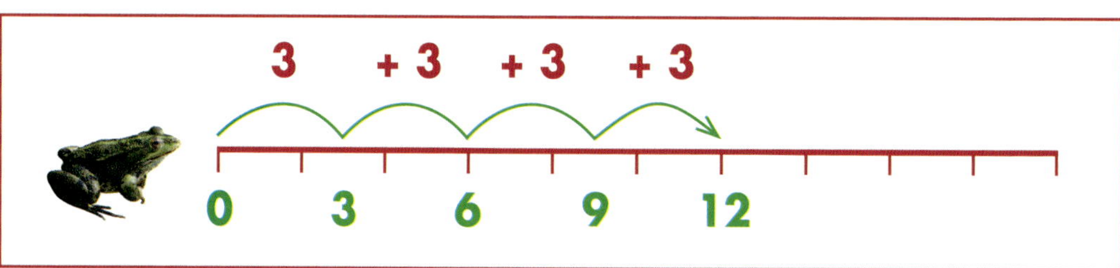

Start at 0 and jump forwards by 3.

4 × 3 = 12

Ava buys 12 rolls.

Try this

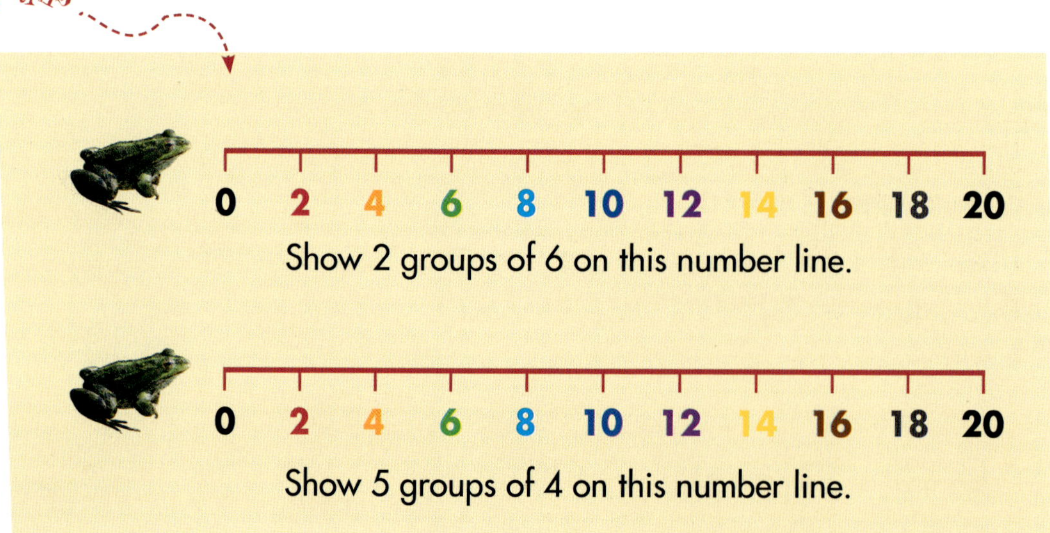

Show 2 groups of 6 on this number line.

Show 5 groups of 4 on this number line.

USE AN ARRAY

An **array** is a set of objects in tidy rows and columns.

It helps you multiply.

The **rows** are all the same length.

The **columns** are all the same length.

It doesn't matter which way up your array goes.

3 rows of 6 = 6 rows of 3

These 2 arrays both show 18 biscuits.

What number sentence matches this array?

There are 4 rows.

Each row has 3 chocolates.

That's 3 + 3 + 3 + 3.

That's 4 × 3 = 12

What number sentence matches this array?

There are 4 rows.

Each row has 4 flowers.

That's 4 + 4 + 4 + 4.

That's 4 × 4 = 16

Draw an array to show 8 rows of 3 dots.

Write the matching number sentence.

GROUPS OF 2

Some things come in pairs.

ears **eyebrows**
eyes **hands** **nostrils**
lips **shoulders**

A pair is a group of 2.

Can you count by 2s quickly with your eyes shut?

0 2 4 6 8 10 12 14 16 18 20

You can draw groups of 2.

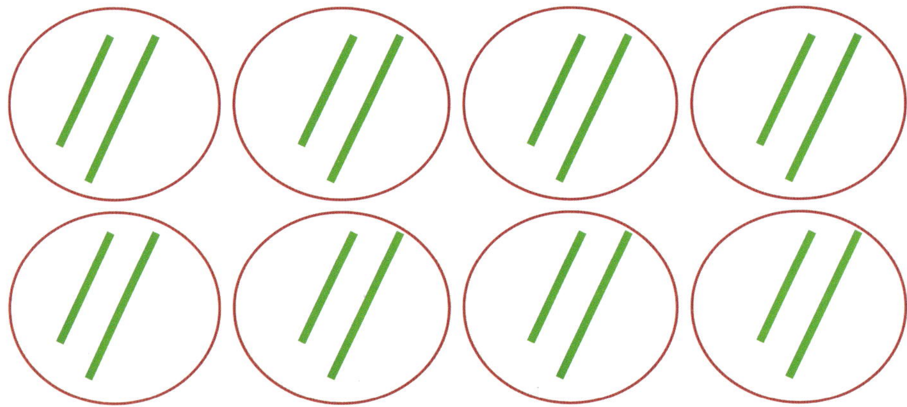

8 groups of 2 make 16

You can put things in a 2s **array**.

3 rows of 2 is the same number as 2 columns of 3

There are 6 still feathers.

You can record groups of 2 on a number line.

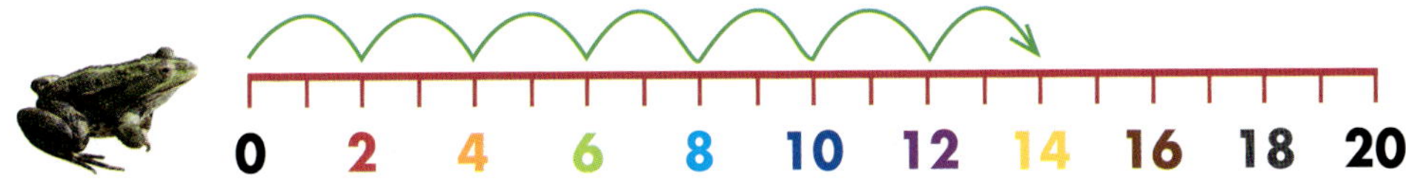

7 jumps of 2 are the same as 14

You can teach your calculator to multiply by 2.

Press 2 + +

Then press =

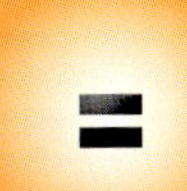

You can solve problems about groups of 2.

How much for 4 bunches of carrots?

$2 +$2 +$2 +$2

4 × $2 = $8

How much change from $10?

10 – 8 = 2

You get $2 change.

0 × 2 = 0
1 × 2 = 2
2 × 2 = 4
3 × 2 = 6
4 × 2 = 8
5 × 2 = 10
6 × 2 = 12
7 × 2 = 14
8 × 2 = 16
9 × 2 = 18
10 × 2 = 20

Write the number sentence to match.

You are a Maths Star if you recall these facts without a mistake

GROUPS OF 5

Some things come in groups of 5.

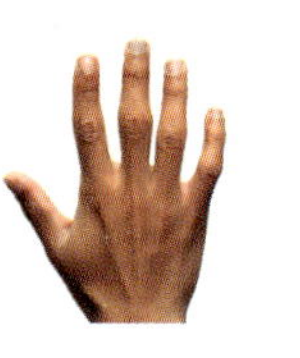
5 fingers

5 leaves

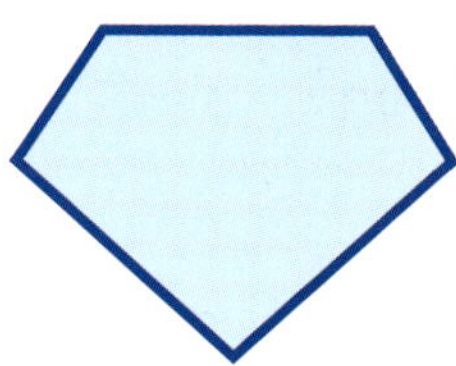
5 sides

Can you count by 5s quickly with your eyes shut?

0 5 10 15 20 25 30 35 40 45 50

You can draw groups of 5.

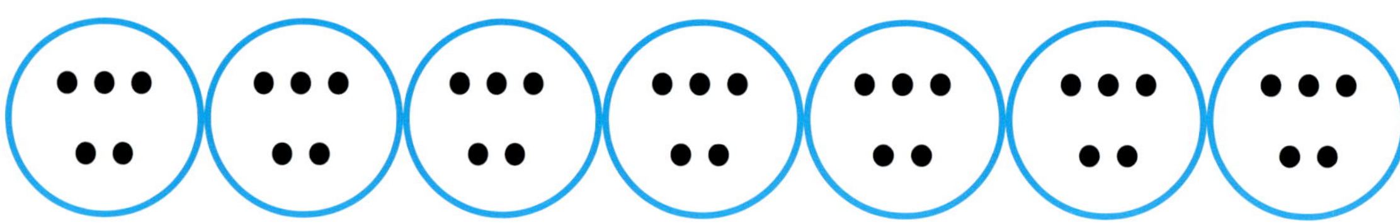

7 groups of 5 is equal to 35

You can put things in a 5s **array**.

4 rows of 5 is the same number as 5 columns of 4

There are 20 people.

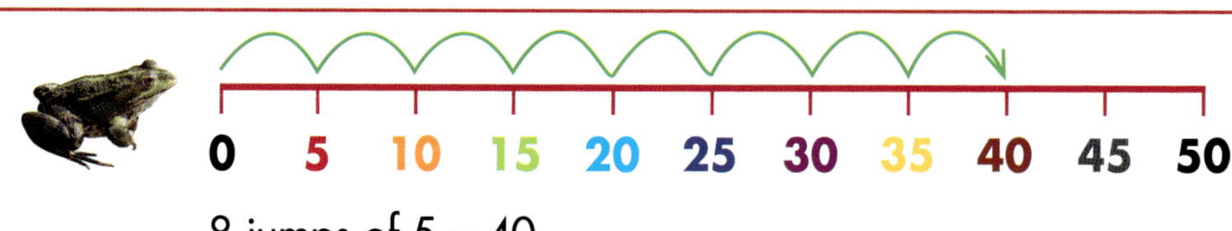

8 jumps of 5 = 40

You can teach your calculator to multiply by 5.

Press 5 + +

Then press =

You can solve problems about groups of 5.

0 × 5 = 0
1 × 5 = 5
2 × 5 = 10
3 × 5 = 15
4 × 5 = 20
5 × 5 = 25
6 × 5 = 30
7 × 5 = 35
8 × 5 = 40
9 × 5 = 45
10 × 5 = 50

How much for 3 bags of apples?

$5 + $5 + $5

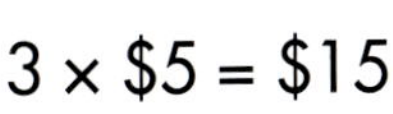

3 × $5 = $15

How much change from $20?

20 − 15 = 5

You get $5 change.

1. Write the number sentence to match.

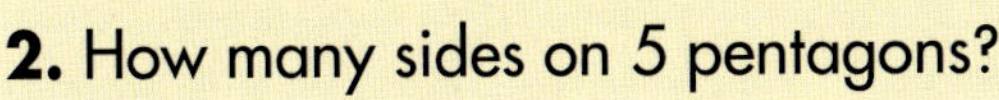

2. How many sides on 5 pentagons?

3. Draw an array to show 10 groups of 5.

Challenge

Write your own story about 8 × 5.

You are a Maths Star if you recall these facts without a mistake

NUMBER & ALGEBRA

GROUPS OF 10

Some things come in groups of 10.

10 legs

10 fingers

10 cents

Can you count by 10s quickly with your eyes shut?

0 10 20 30 40 50 60 70 80 90 100

You can draw groups of 10.

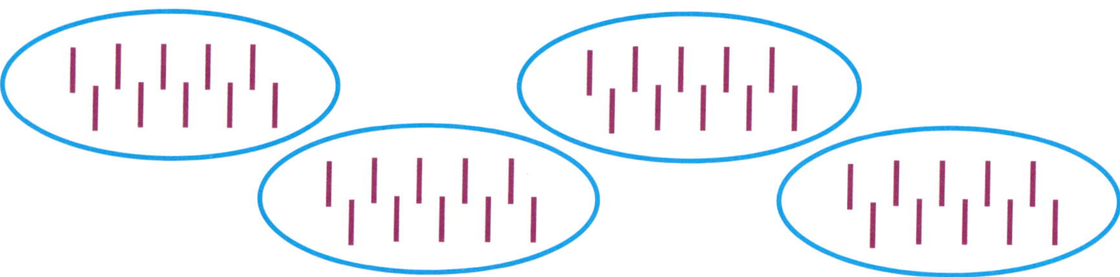

4 groups of 10 sticks

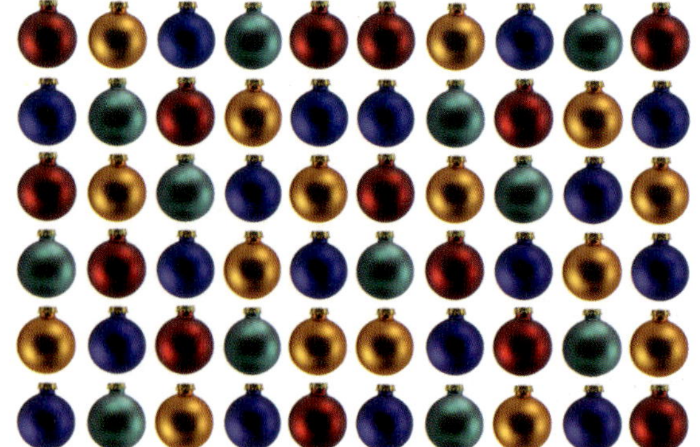

You can put things in a 10s **array**.

6 rows of 10 are the same as 10 columns of 6

There are 60 balls.

You can record groups of 10 on a number line.

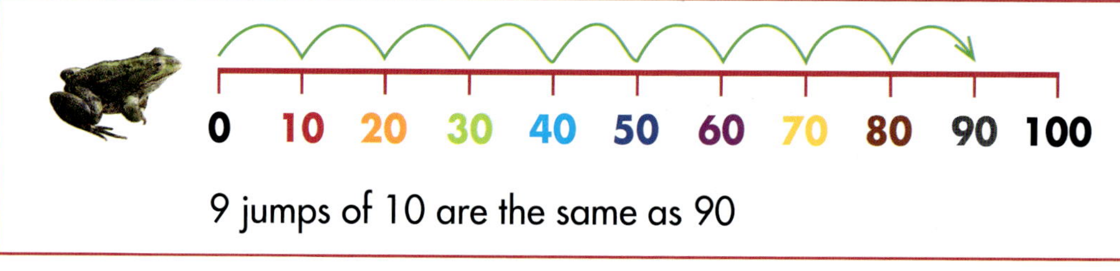

9 jumps of 10 are the same as 90

You can teach your calculator to multiply by 10.

Press 1 0 + + Then press =

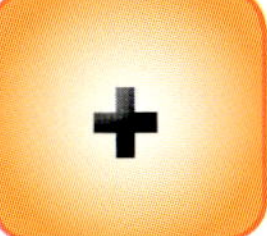

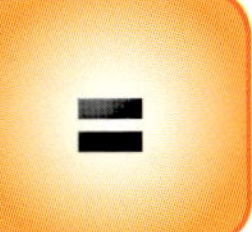

You can solve problems about groups of 10.

How much for 5 bears?

$10 +$10 +$10 +$10 +$10

$5 \times \$10 = \$50$

You can pay for that with one $50 note.

Try this

Try to complete each wheel in less than 1 minute.

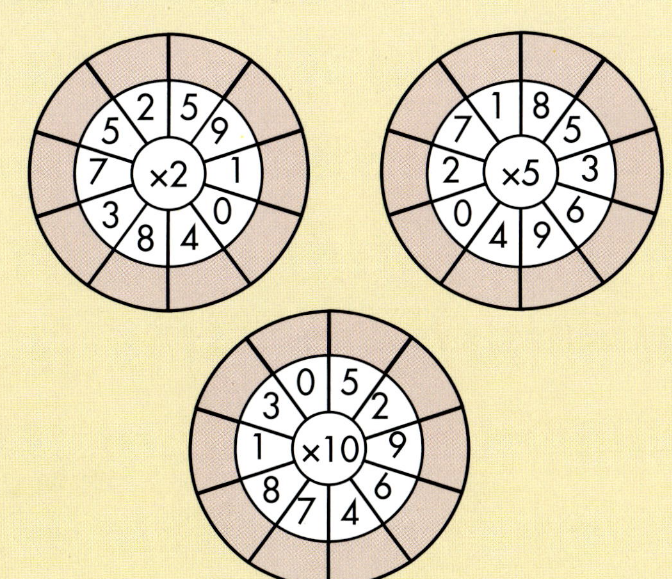

0 × 10 = 0

1 × 10 = 10

2 × 10 = 20

3 × 10 = 30

4 × 10 = 40

5 × 10 = 50

6 × 10 = 60

7 × 10 = 70

8 × 10 = 80

9 × 10 = 90

10 × 10 = 100

You are a Maths Star if you recall these facts without a mistake

HOW TO DIVIDE

Division is repeated subtraction

Division is a fast way to subtract.

It only works if you take away the same number each time.

Here are 18 fish.

You divide when you make equal groups

How many groups of 3 fish?

Draw a line around each group of 3.

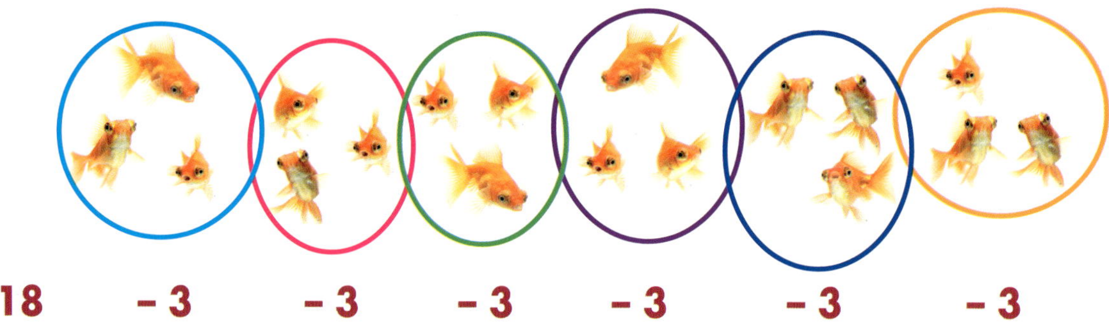

18 - 3 - 3 - 3 - 3 - 3 - 3

You can take away 6 groups of 3.

Are there any fish left over? No.

There are 6 groups of 3 fish.

This is what it looks like on a number line.

Start at 18 and jump back by 3.

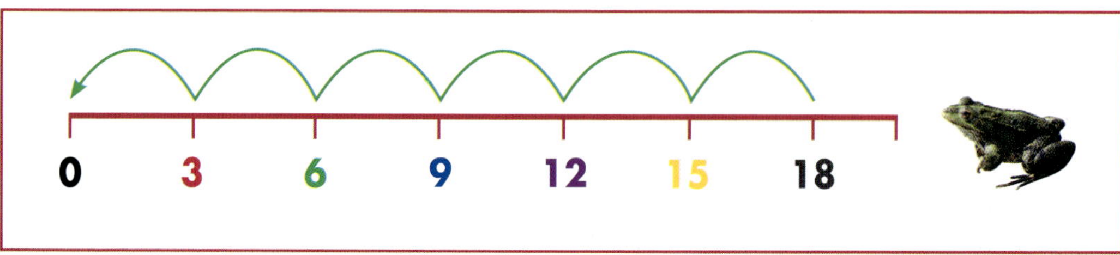

Sometimes you have leftovers.

Lily has 19 bananas.

How many groups of 4 bananas?

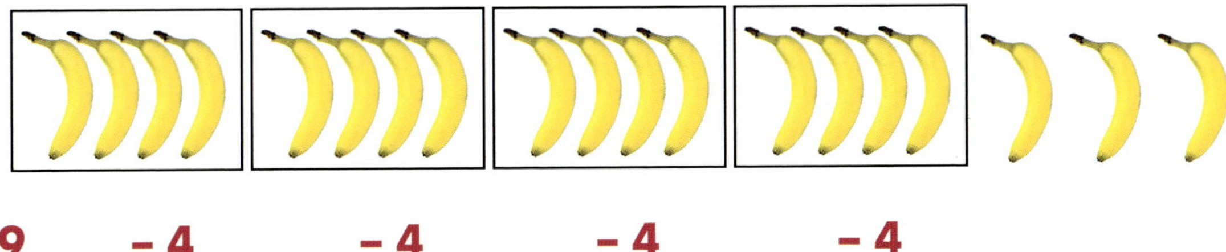

19 **- 4** **- 4** **- 4** **- 4**

She takes away 4 groups of 4.

Are there any bananas left over?

Yes, 3 bananas are left.

So Lily has 4 groups of 4 and 3 extras.

If Lily had 1 more banana she would have 5 groups of 4.

This is what it looks like as a drawing.

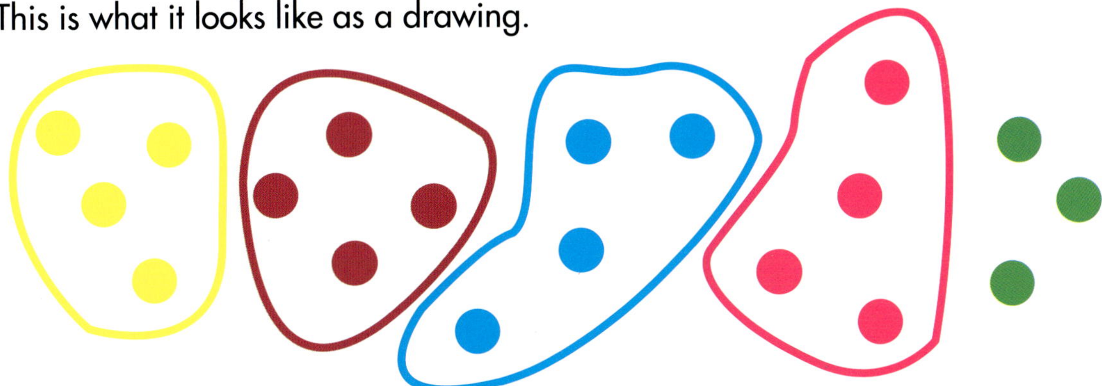

This is what it looks like on a number line.

Start at 19 and jump back by 4.

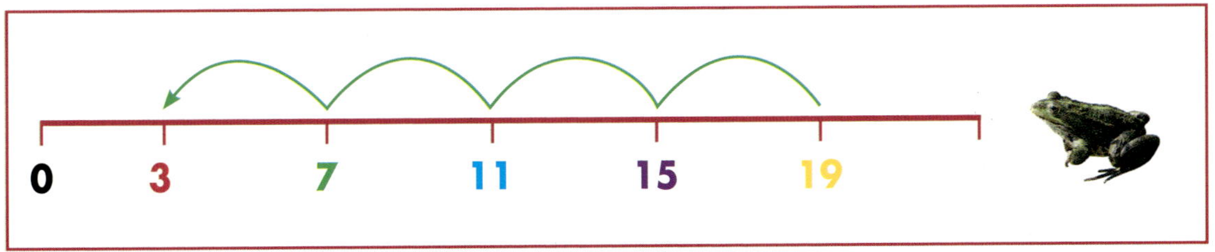

4 groups of 4 and 3 extras make 19.

HOW TO DIVIDE (continued)

Six friends won $60.

How much money do they each get?

They all want the same share.

Six groups of what makes $60?

6 groups of ☐ makes 60.

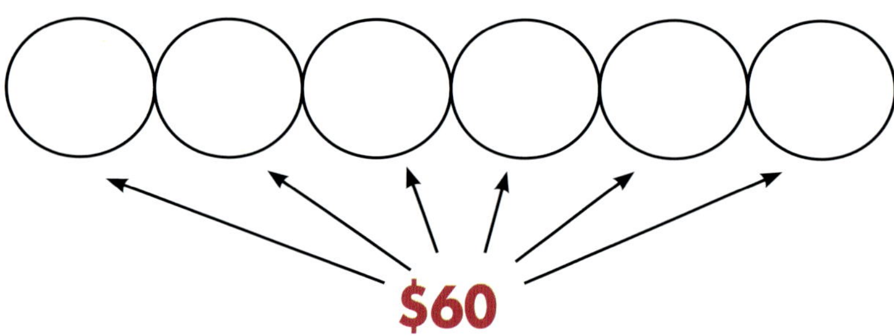

Use guess and check to find out.

If they got **$5** each, that's 6 groups of 5 or $30.

$60 – $30 = $30.

There is still $30 left to share.

That's another **$5** each.

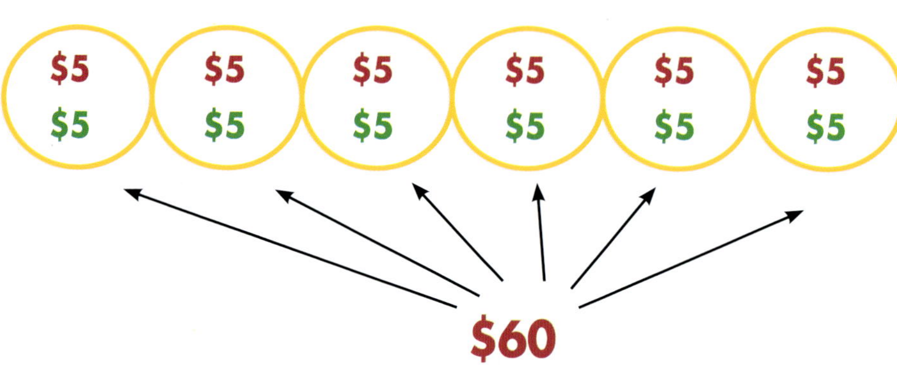

So each person gets **$5** + **$5** = $10.

Six groups of $10 makes $60.

6 groups of 10 makes 60.

Share 10 eggs between 2 boys.

That's 5 eggs each.

Sometimes you have leftovers.

Three children share 14 books.

How many books each?

3 groups of ☐ make 14

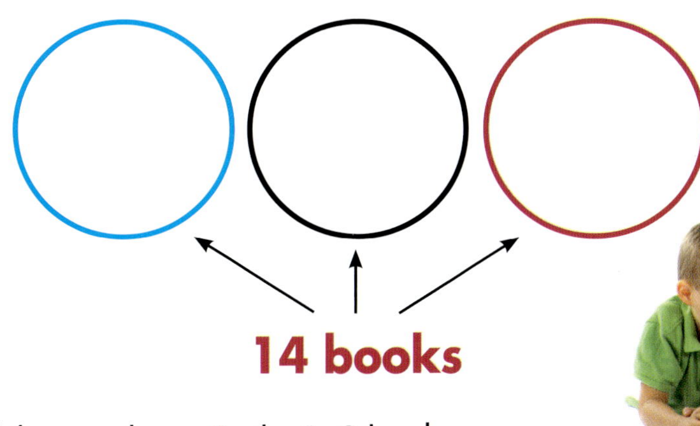

If they each get 1, that's 3 books.

If they each get another 1, that's 6 books.

If they each get another 1, that's 9 books.

If they each get another 1, that's 12 books.

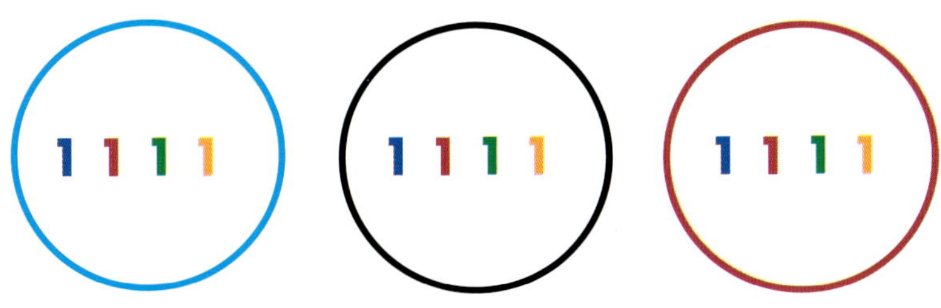

If they have 1 more book there will be 3 groups of 5

That's 4 books each with 2 books left over.

That's 3 groups of 4 books and 2 extras.

NUMBER & ALGEBRA

MORE EXAMPLES OF GROUPS AND SHARES

How many groups of 2 shells?

When you divide you share or group

11 groups of 2 with no extra shells.

How many groups of 5 flowers?

☐ groups of 5 flowers.

3 groups of 5 with 1 left over.

Share 18 bones between 5 dogs.

That's 3 bones each with 3 left over.

Share 13 cakes between 10 girls.

That's 1 cake each with 3 extras.

How many groups of 3 beetles?

____

FRACTIONS

Numbers like 2, 54, 109 and 876 are **whole numbers**.

Not all numbers are whole numbers.

Some numbers are parts of a whole.

When you divide a whole object into equal size parts, you create **fractions**.

A whole strawberry

CREATE HALVES

2 equal parts
2 **halves**
Each part is one half.

These are not 2 equal parts.
These are not 2 halves.

Here are some more halves:

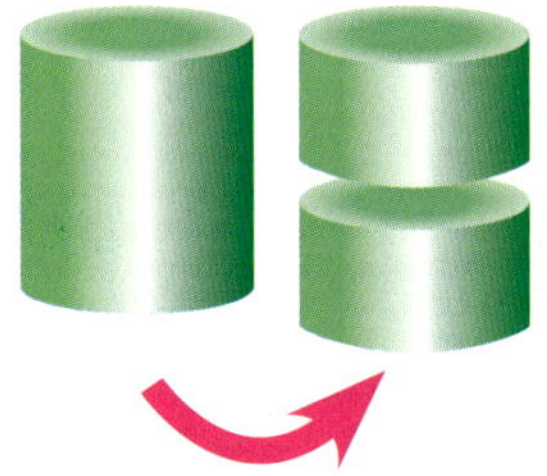

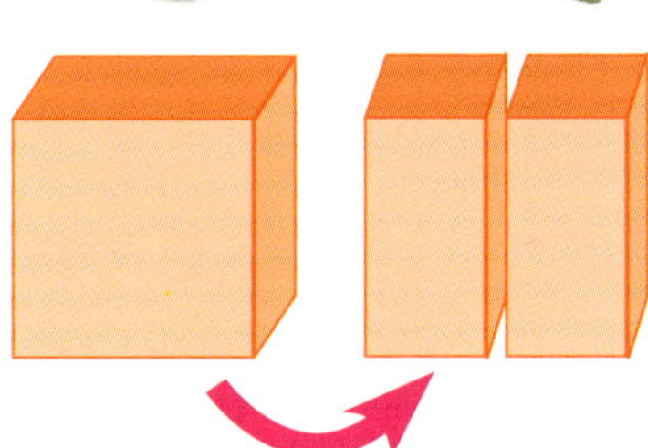

You create halves when you fold some paper shapes into 2 equal size parts.

A paper square

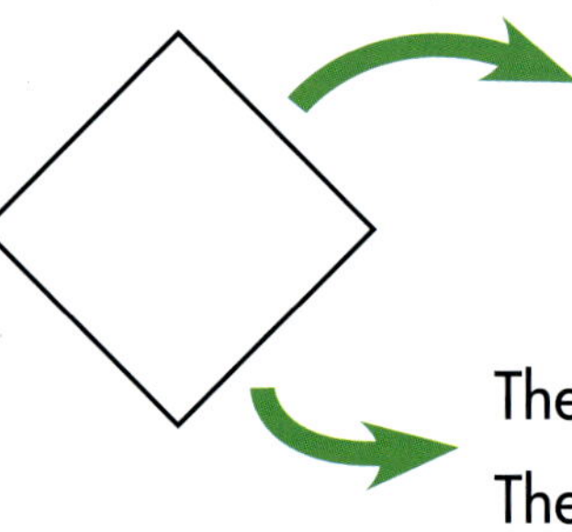

2 equal parts

2 halves

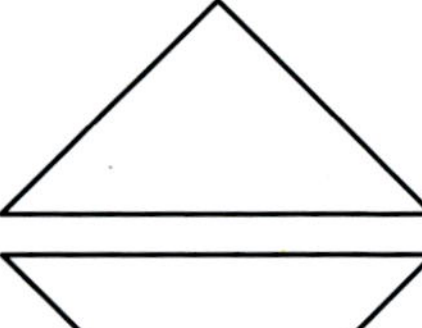

These are not 2 equal parts.

These are not 2 halves.

Here are some more halves:

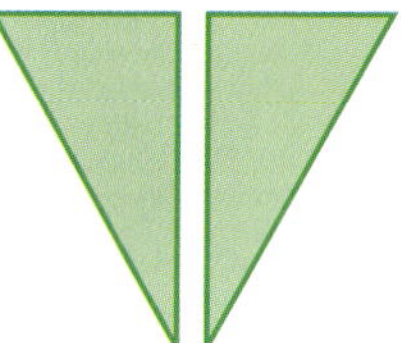

You create halves when you fold a length of rope or string into 2 equal lengths.

A skipping rope

2 equal lengths

2 halves

These are not 2 equal lengths.

You create halves when you share objects into 2 equal groups.

1 is half of 2

3 is half of 6

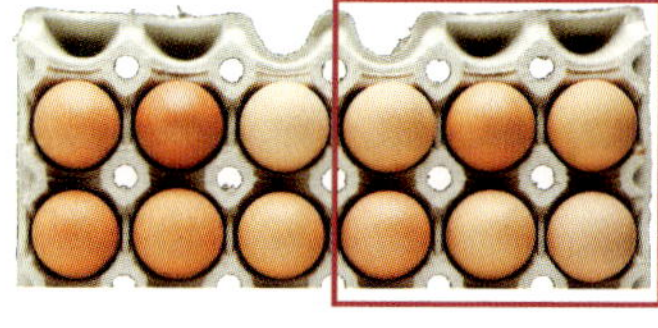

6 is half of 12

These are not 2 equal groups.

These are not halves.

Mathematicians write $\frac{1}{2}$ to represent 1 half.

2 halves are exactly the same as one whole.

2 halves

1 whole

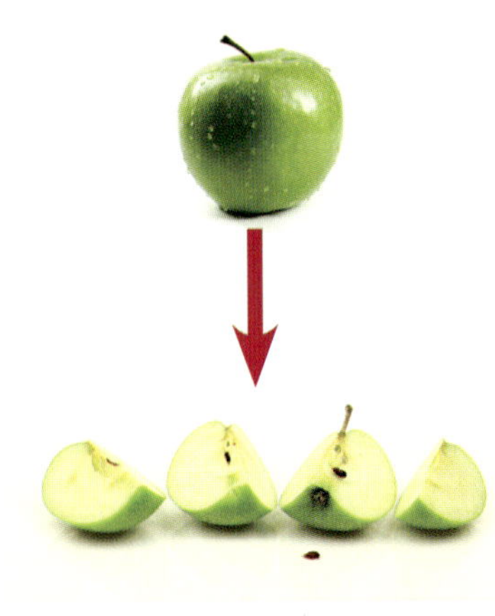

CREATE QUARTERS

You create quarters when you divide a whole object into 4 equal parts.

A **quarter** is one of 4 equal parts.

It is also called a **fourth**.

Each quarter is half of a half.

If you chop a half into 2 equal parts you get quarters.

1 whole  2 halves  4 quarters 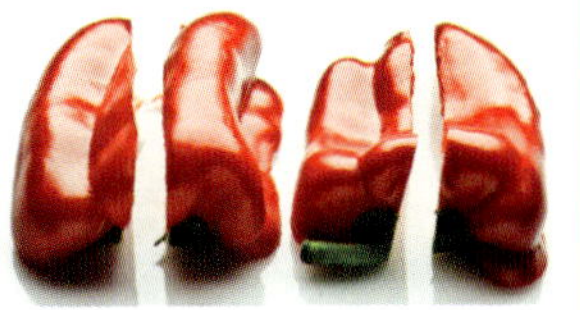

There are 4 equal parts.

Each part is one quarter.

You can add quarters together.

1 quarter is smaller than one half.

2 quarters are the same size as 1 half.

4 quarters are the same size as 1 whole.

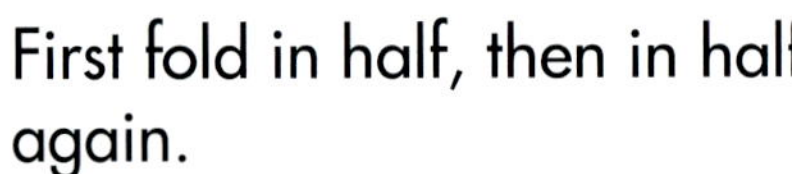

You create quarters when you fold some paper shapes into 4 equal size parts.

First fold in half, then in half again.

Open out to reveal your 4 quarters.

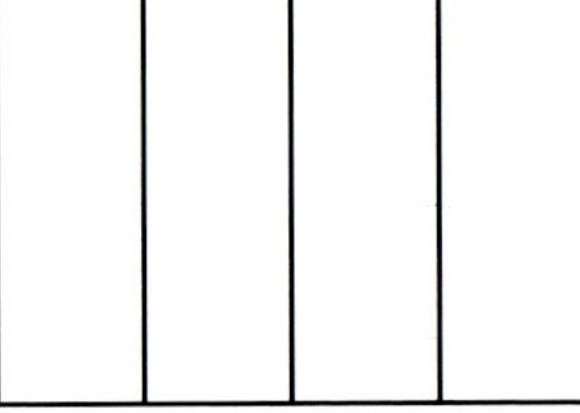

You create quarters when you fold and cut a length into 4 equal size lengths.

2 halves

4 quarters

Mathematicians write quarters like this:

You create quarters when you share objects into 4 equal size groups.

1 is a quarter of **4**.

2 is a quarter of **8**.

3 is quarter of **12**.

These are not 4 equal groups.

These are not quarters.

Cut a length of paper as long as your body.

Fold this length in half. Fold it in half again.

Use this to measure lengths around you.

Find something as long as:

- $\frac{1}{4}$ of your height
- $\frac{2}{4}$ of your height
- $\frac{3}{4}$ of your height.

Try this

NUMBER & ALGEBRA

An eighth is one of 8 equal parts

EIGHTHS

You create **eighths** when you divide a whole object into 8 equal parts.

An eighth is smaller than a quarter.

Each eighth is half of a quarter.

If you chop a quarter into 2 equal parts you get one eighth.

1 whole

2 halves

4 quarters

8 eighths

You can add eighths together.

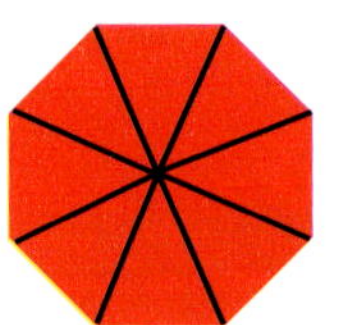
$\frac{8}{8}$ of a regular octagon

a regular octagon

8 eighths are exactly the same size as 1 whole.

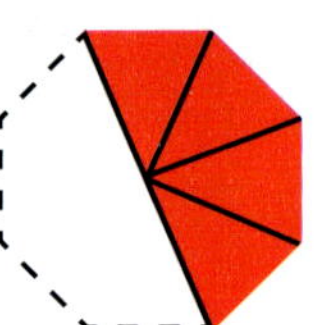
$\frac{4}{8}$ of a regular octagon

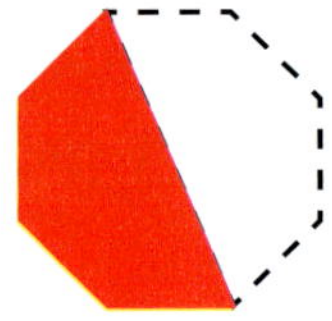
$\frac{1}{2}$ of a regular octagon

4 eighths are exactly the same size as 1 half.

You create eighths when you fold or cut a length into 8 equal lengths.

These are not 8 equal lengths.

These are not eighths.

You create eighths when you share objects into 8 equal size groups.

1 is an eighth of 8

2 is an eighth of 16

3 is an eighth of 24

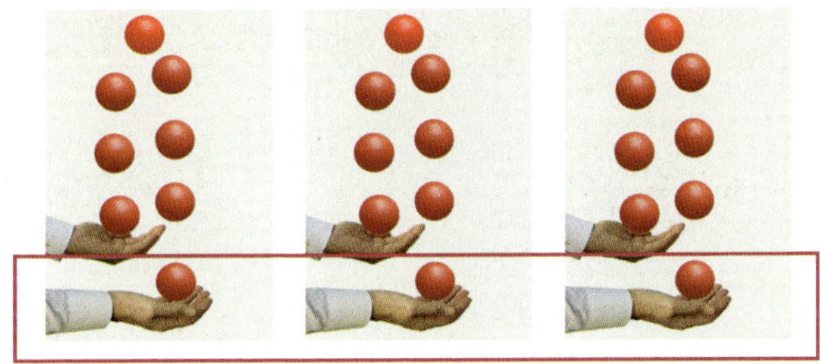

These are not 8 equal groups. These are not eighths.

Mathematicians write 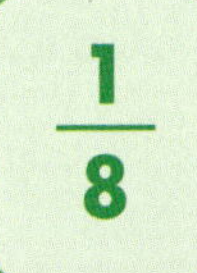 $\frac{1}{8}$ to represent 1 eighth.

1 whole							
1 half				1 half			
1 quarter		1 quarter		1 quarter		1 quarter	
1 eighth	1 eighth	1 eighth	1 eighth	1 eighth	1 eighth	1 eighth	1 eighth

You need a plastic knife and a ball of playdough.
Explain to a friend how to cut the ball into eighths.

NUMBER & ALGEBRA

MONEY

Australia uses **decimal currency**.

You use 6 coins and 5 notes.

These are based on groups of 10 and 100.

Some people say you won't use coins or notes in future.

SILVER COINS

5 cent coin

5c

This is the smallest Australian coin.

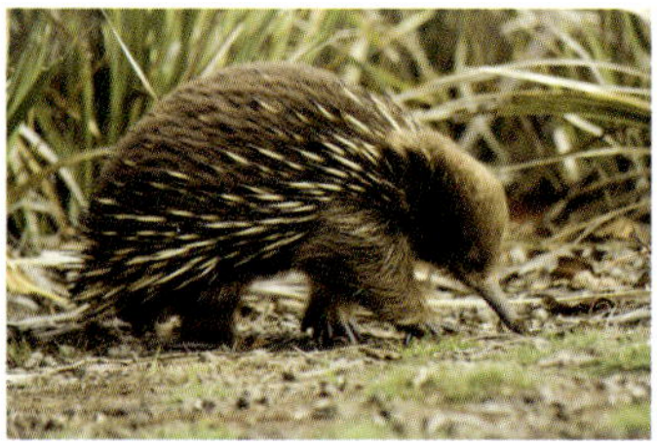

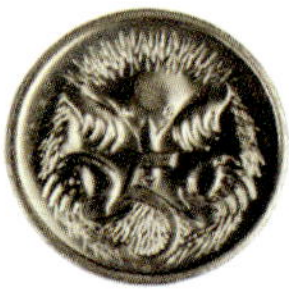

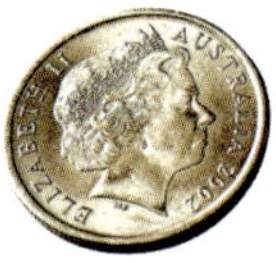

front an echidna **reverse** Queen Elizabeth II

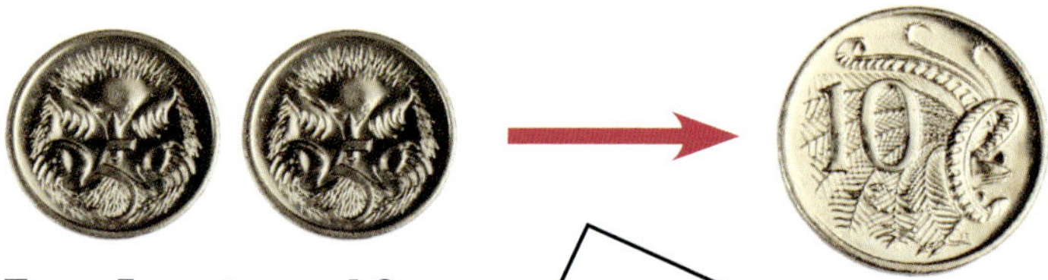

Two 5c coins = 10c

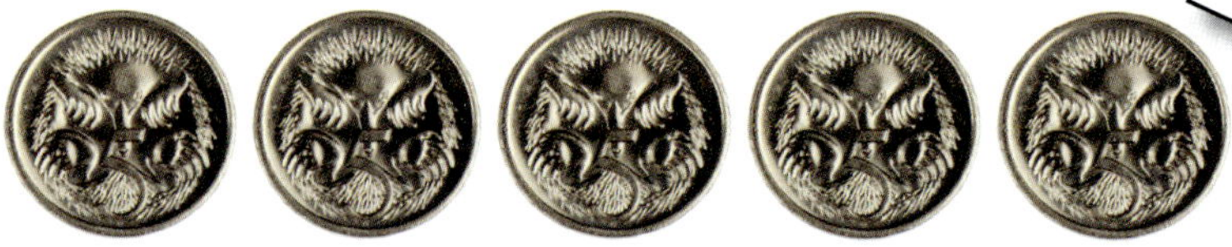

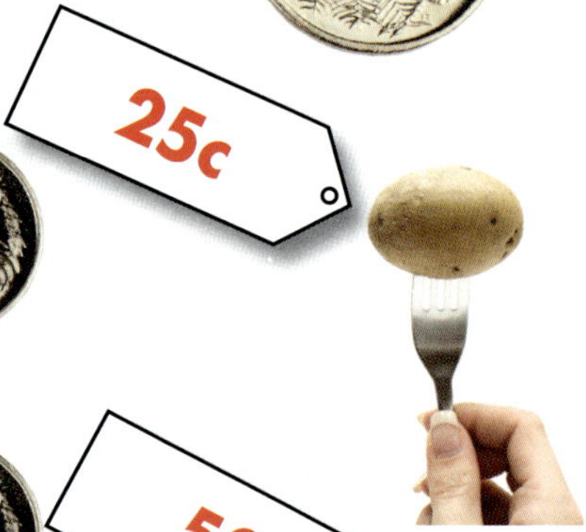

25c

Five 5c coins = 25c

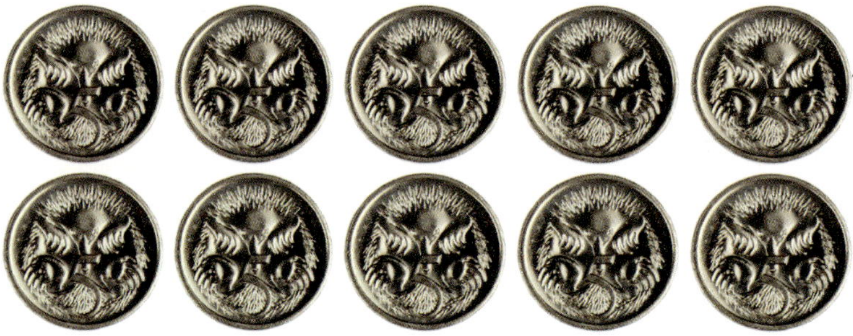

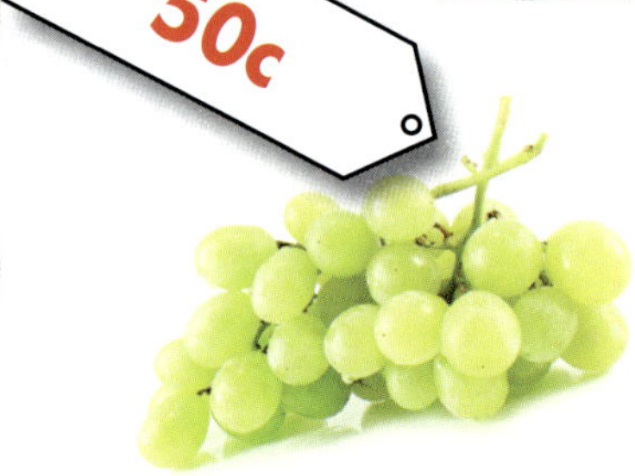

50c

Ten 5c coins = 50c

10 cent coin

This coin is larger than a 5c coin.

front a lyre bird

reverse Queen Elizabeth II

Two 10c coins = 20c

Five 10c coins = 50c

Ten 10c coins = $1

You can count these 10 coins like this:

10 20 30 40 50 60 70 80 90 100 cents

Or you can count these coins this way:

10c 20c 30c 40c 50c 60c 70c 80c 90c $1

20 cent coin

This coin is larger than a 10c coin.

front a platypus **reverse** Queen Elizabeth II

Two 20c coins = 40c

Five 20c coins = $1

Ten 20c coins = $2

You can count these 10 coins like this:

20 40 60 80 100 120 140 160 180 200 cents

Or you can count these 10 coins this way:

20c 40c 60c 80c $1.00 $1.20 $1.40 $1.60 $1.80 $2

50 cent coin

This is the largest Australian coin.

front
a kangaroo and an emu

reverse
Queen Elizabeth II

$1

Internet $2.50 per hour

Two 50c coins = $1

Five 50c coins = $2 and 50c

Ten 50c coins = $5

You can count these 10 coins like this:

50 100 150 200 250 300 350 400 450 500 cents

Or you can count these 10 coins this way:

50c $1 $1.50 $2 $2.50 $3 $3.50 $4 $4.50 $5

You write $2 and 50c as $2.50.

You read the dot as 'and'.

GOLD COINS

100 cent coin

front a mob of kangaroos

reverse Queen Elizabeth II

Two $1 coins = $2

Five $1 coins = $5

Ten $1 coins = $10

You can count these 10 coins like this:

1 2 3 4 5 6 7 8 9 10 dollars

Or you can count these 10 coins this way:

$1 $2 $3 $4 $5 $6 $7 $8 $9 $10

200 cent coin

This coin is smaller than a $1 coin.

2 dollars
$2

front an aboriginal hunter **reverse** Queen Elizabeth II

Two $2 coins = $4

Five $2 coins = $10

Ten $2 coins = $20

You can count these 10 coins like this:

2 4 6 8 10 12 14 16 18 20 dollars

Or you can count these 10 coins this way:

$2 $4 $6 $8 $10 $12 $14 $16 $18 $20

COMBINING COINS

There are many ways to add coin values to make a total price.

Jonno buys an iceblock using these coins.

He counts the coins like this:

5 10 20 30 50 $1

He adds them up as he counts.

The iceblock costs $1.

Sally buys a drink using these coins.

10c

+ 20c

+ 20c

+ 50c

+ $1

She counts the coins like this:

10 30 50 $1 $2

She adds them up as she counts.

The drink costs $2.

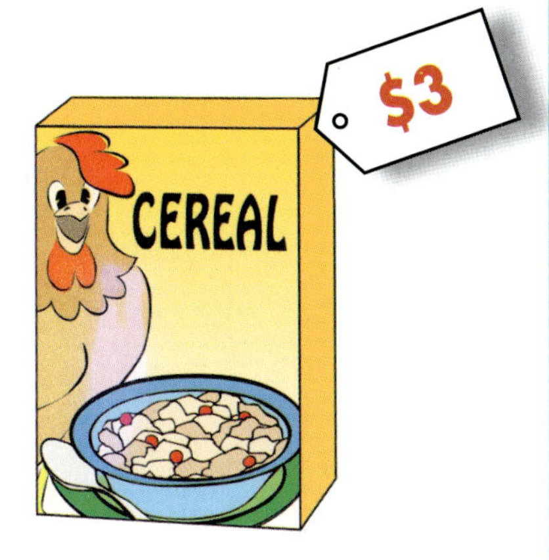

Lee buys this box of cereal with 6 coins.

10c + 20c + 20c + ☐ + $1 + $1

Which coin is missing?

10c + 20c + 20c + $1 + $1 = $2.50

Another 50c makes $3.

A 50c coin is missing.

Here are 3 different ways to pay this bus fare using up to 6 coins.

$1 + $1 + $1 + $1 +$1

50c + 50c + $1 + $1 + $2

10c + 20c + 20c + 50c + $2 + $2

Oliver has three dollars and 75 cents in his pocket.

What could these coins be?

____

____

Challenge

Lara buys three $2 pencils plus two $9 books.

What change does she get from $50?

____

____

____

____

Five dollars

PLASTIC NOTES

This is the smallest note.

$5 is worth the same as five $1 coins

front

Parliament House

colour → mauve

reverse

Queen Elizabeth II

$5 per pair

How much for 8 pairs?

8 × $5 = $40

How much change from $50?

$50 – $40 = $10

Two $5 notes = $10

Five $5 notes = $25

Ten $5 notes = $50

Ten dollars

colour → blue

front

Banjo Paterson

reverse

Mary Gilmore

$10 is worth the same as two $5 notes

How much for 6 toys?

6 × $10 = $60

How much change from $100?

$100 – $60 = $40

Two $10 notes = $20

Try this

How many different ways can you make $10 using coins?

Five $10 notes = $50

Ten $10 notes = $100

Twenty dollars

front

Mary Reibey

reverse

John Flynn

$20 is worth the same as ten $2 coins

Two $20 notes = $40

Five $20 notes = $100

Ten $20 notes = $200

How much for 4 budgies?

4 × $20 = $80

How much change from $100?

$100 – $80 = $20

Challenge

Fifi's vet bill includes a check-up plus $60 for food and $120 for medicine.

Her owner paid $400 and got $80 change.

How much was the check-up?

____

Fifty dollars

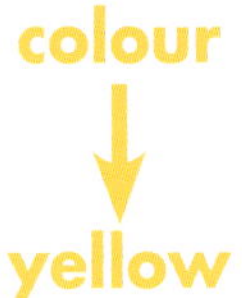

front

David Unaipon

reverse

Edith Cowan

$50 is worth the same as twenty-five $2 coins

Two $50 notes = $100

Five $50 notes = $250

Ten $50 notes = $500

$50 is worth the same as two $20 notes and one $10 note

How much for 3 umbrellas?

$50 + $50 + $50 = $150

How much change from $200?

$200 – $150 = $50

One hundred dollars

$100

This is the largest note.

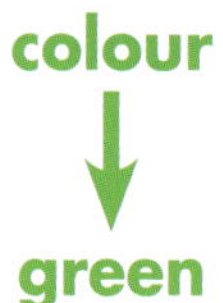

front

Nellie Melba

reverse

John Monash

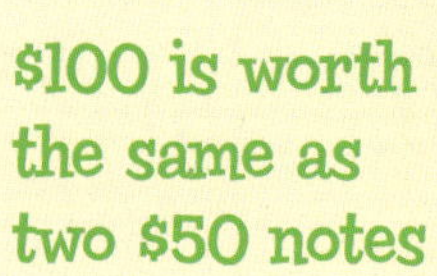

Two $100 notes = $200

$100 for 4 haircuts

How much for 1 haircut?

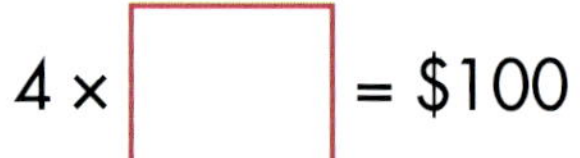

4 × ☐ = $100

Divide $100 into 4 equal parts.

$100 = $25 + $25 + $25 + $25

One haircut costs $25.

Five $100 notes = $500

Ten $100 notes = $1000

Australia does not have a $1000 note

Try this

How many different ways can you make $50 using just notes?

____

Challenge

Find a shopping catalogue.

You want to buy 6 things for no more than $100.

What will you buy?

PATTERNS

You can:

make a pattern using a rule

describe a pattern to a friend

continue a pattern

predict the next part

discover a missing part

record a pattern

MAKE A PATTERN WITH OBJECTS

This is a colour pattern.

Look carefully at the position of each colour.

The next part of this pattern will be 

You can record it like this:      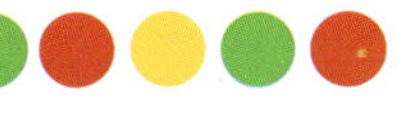

This is a shape pattern.

Which shape is missing?

Look carefully at the position of each shape.

This is a size pattern. Look carefully at the size of each object.

The next parts in this pattern will be  then  then 

MAKE A PATTERN WITH OBJECTS (continued)

?

This is a colour and shape pattern.

Which shape is missing?

Look carefully at the position of each shape.

This is a colour and number pattern.

The next part of this pattern will be

You can record it like this:

Or you can record it with numbers:

123123123123123123123

This is a position pattern. Look carefully at the position of each fan.

The next part of this pattern will be

Try this

Draw the missing shapes in this pattern.

Draw the missing beads in this pattern.

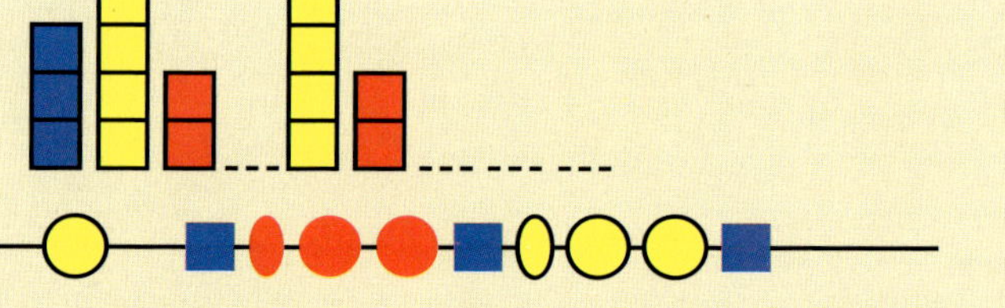

'ADD A NUMBER' PATTERNS

These are all odd numbers.

You add 2 each time.

11 and **13** are the next 2 numbers in this pattern.

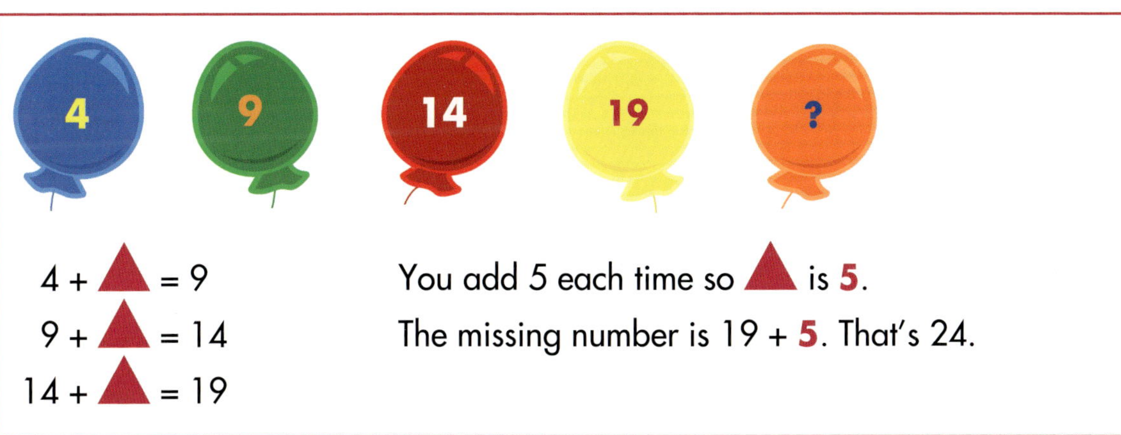

4 + ▲ = 9
9 + ▲ = 14
14 + ▲ = 19

You add 5 each time so ▲ is **5**.

The missing number is 19 + **5**. That's 24.

Which numbers are missing?

Try to work out what to add each time.

11 + ◆ = 14
14 + ◆ = 17

You add 3 each time so ◆ is **3**.

The missing numbers are **8** and **20**.

The next number will be 23.

What are the next 3 numbers in each pattern?

RAINBOW PATTERN

Start at any number.

Write numbers in a pattern.

Join them up like a rainbow.

Like magic, each pair has the same total.

This is an 'add 1' pattern.

4 + 17 = 21

5 + 16 = 21

6 + 15 = 21

7 + 14 = 21

8 + 13 = 21

This works for any addition pattern.

This is an 'add 5' pattern.

0 + 35 = 35

5 + 30 = 35

10 + 25 = 35

15 + 20 = 35

If one number remains in the middle, just double it.

'SUBTRACT A NUMBER' PATTERNS

You take away 2 each time.

The missing number is 10 – 2. That's 8.
The next number after that will be 6.

Work out what to subtract each time.

27 – ◯ = 22   22 – ◯ = 17   17 – ◯ = 12

You subtract 5 so ◯ = 5.

The next numbers are **7** and **2**.

Which numbers are missing?

Try to work out what to subtract each time.

67 – ⬠ = 57   57 – ⬠ = 47

You subtract 10 so ⬠ = 10

The missing numbers are **37** and **17**.

Challenge

Describe 4 different patterns you see in this 100s chart.

0	1	2	3	4	5	6	7	8	9
10	11	12	13	14	15	16	17	18	19
20	21	22	23	24	25	26	27	28	29
30	31	32	33	34	35	36	37	38	39
40	41	42	43	44	45	46	47	48	49
50	51	52	53	54	55	56	57	58	59
60	61	62	63	64	65	66	67	68	69
70	71	72	73	74	75	76	77	78	79
80	81	82	83	84	85	86	87	88	89
90	91	92	93	94	95	96	97	98	99

LENGTH

Let's find out about the sizes of things

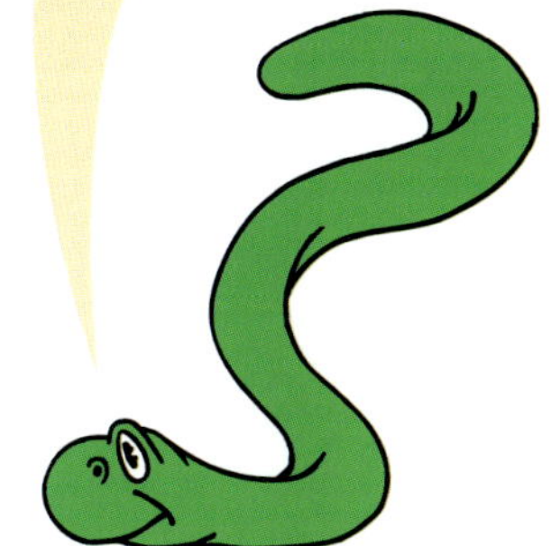

These words help you talk about **length**:

long **tall** **short** **low** **narrow** **wide** **high** **far** **near**

a long body

short legs

a tall giraffe

a high jump

a narrow path

a wide sofa

COMPARING STRAIGHT LENGTHS

It is easy to compare the length of **straight** lines or edges.

Is one shorter than, the same length as, or longer than the other?

Compare 2 lengths by matching one end.

You can also compare lengths by putting things in order from shortest to tallest.

Shortest to tallest

Tallest to shortest

These carrots are **not** in order by length.

The 1st carrot is the shortest.

The 4th carrot is the longest.

The last carrot is almost as long as the 4th carrot.

COMPARING CURVED OR CROOKED LENGTHS

a curved path

a zig zag wall

It is not so easy to measure the length of curved or crooked lines or edges

How will you measure the length of this banana?

It is longer than this line

Try matching the curved length with string.

You can now pull the string straight.

The banana is as long as this string.

Try this string method when comparing lengths that can't be placed directly together.

Jess measures around her hoop with a ribbon.

The ribbon is longer than her body.

Try this

Draw your own curved or crooked line.

Match this length with string.

Find 2 more things that are about as long as your line.

MEASURING WITH LENGTH UNITS

You can use small equal length **units** to measure and compare longer lengths.

There are many different units you can use.

This carrot is just a bit shorter than 3 bricks.

Jack is just a bit longer than 6 shoes.

Each small unit is exactly the same length.

These scissors are as long as 4 paper clips.

Use units that are exactly the same length

Make sure you start measuring exactly at one end.

Repeat each unit with **no gaps** and **no overlaps**.

These paper clips have no gaps or overlaps.

Use longer units to measure very long lengths.

These shoes have gaps between them.

They do not start at one end.

How many cars fit along one side of the carpark?

Make your own body measure.

Cut a strip of paper as tall as your height.

Use this to find things around your house that are shorter than, exactly the same length as, or longer than your body length.

Try this

Challenge

Make a footstep measurer.

Cut a strip of paper as long as 10 footsteps.

Mark off where each footstep reaches.

How many footsteps long is your bedroom?

Estimate first, then check.

MEASURING WITH PARTS OF A UNIT

Sometimes there are bits left over.

You can talk about these using fraction names.

Here are some half units:

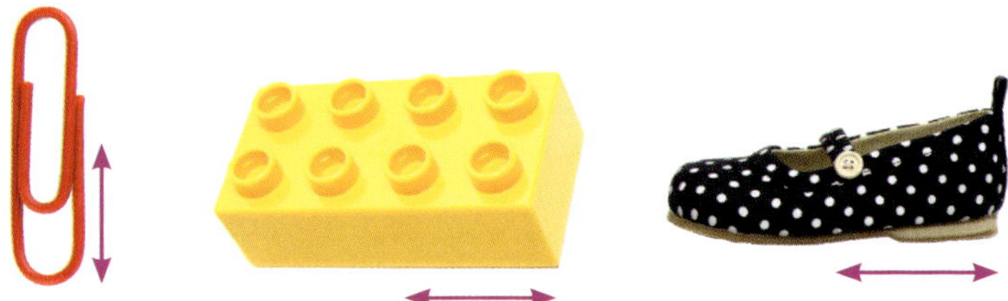

Here are some quarter units:

Imagine how much is left over when you measure.

Imagine it is a fraction of the whole unit.

Use half and quarter units to describe lengths more accurately.

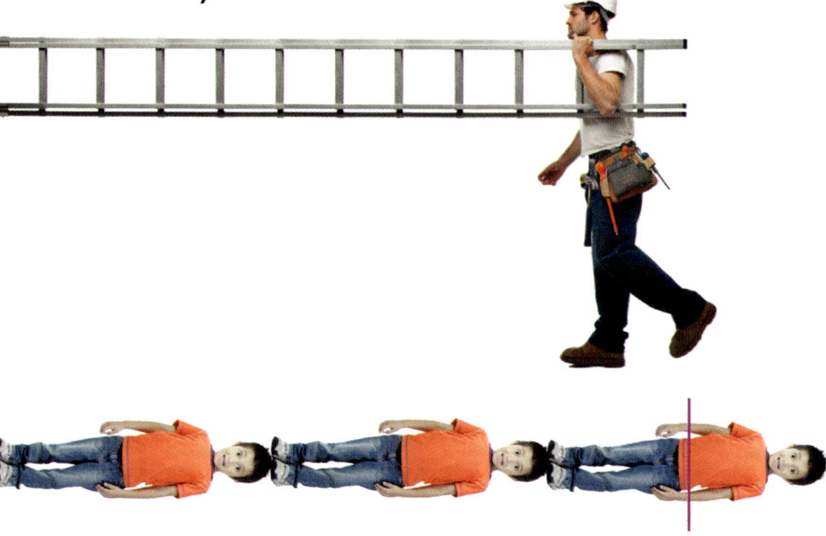

This ladder is as long as 2-and-a-half Jacks.

MEASURING WITH METRES

It is not precise to measure length with units like shoes, body or car lengths.

Shoes, bodies and cars can be long or short.

You can end up with different numbers.

Most people in the world agree to measure length using a **standard unit**.

The standard unit of length is a **metre**.

One metre is as long as 4 Maths Guides.

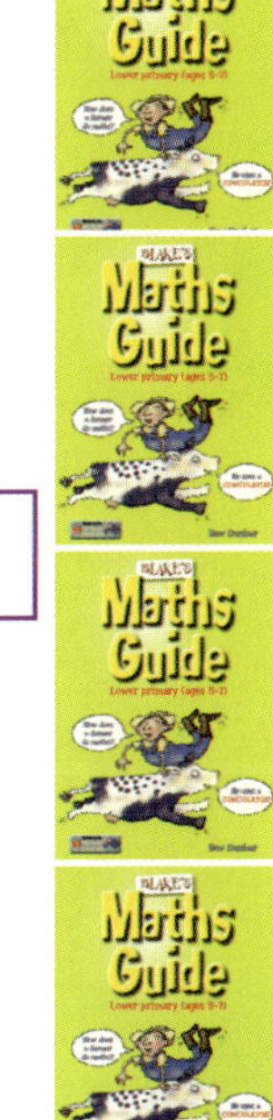

You use a metre ruler or a tape measure to measure length in metres.

The symbol for metre is m.

No matter where in the world you measure, these lengths will always be the same.

When you think you know how to measure with metres, estimate first, before you check.

Try this

Use a measuring tape to find a body metre.

Stretch out your arms.

Ask a friend to measure exactly one metre from one fingertip along your arm.

Remember this length. Use it to measure metres around you.

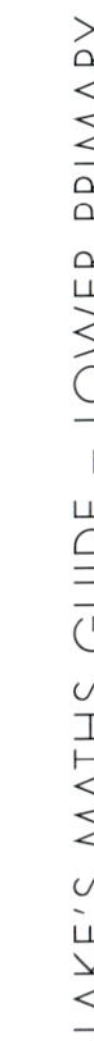

A giraffe can grow taller than 5 metres.

MEASURING WITH PARTS OF A METRE

When you measure with a tape, sometimes there are bits left over.

You can talk about these bits using fraction names.

The longest blue whale is about 32 and ¼ m long.

2 halves

½ metre ½ metre

1 metre

4 quarters

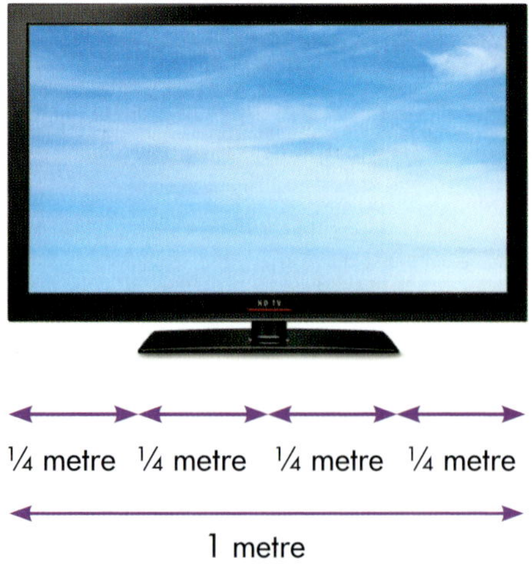

¼ metre ¼ metre ¼ metre ¼ metre

1 metre

You can fold a metre tape in half to check your estimate

Imagine how much is left over when you measure.

Use half and quarter metres to describe lengths more accurately.

Practise estimating ½ metre and ¼ metre lengths.

MEASURING WITH CENTIMETRES

A metre is a large unit of length.

Lengths shorter than a metre are measured with short, equal length units called **centimetres**.

cm

The symbol for centimetre is cm.

Centimetres are also an accurate way to measure bits left over.

You use a centimetre ruler or a tape measure.

Line up 0 on your ruler to the start of the object.

Read the number where the length finishes.

This girl is **not** measuring correctly.

She does not have the tape measure starting at 0.

The world's longest caterpillar is about 12 centimetres.

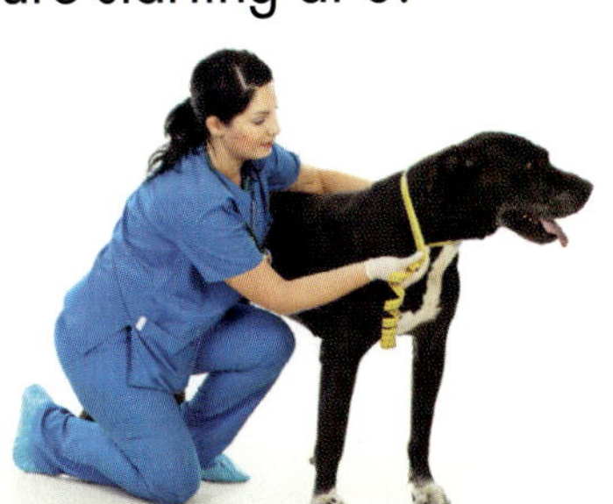

Use a soft tape measure to measure curved lengths.

When you think you know how to measure with centimetres, estimate first, before you check.

Try this

Look at this man measuring the fish.

Explain what he is doing wrong.

The average Chihuahua is about 20 cm tall.

0 1 2 3 4 5 6 7 8 9 10 11 12 13 14 15 16 17 18 19 20 21 22 cm

MEASUREMENT & GEOMETRY

AREA

Area is a measure of how much **surface** a flat object covers.

These words help you talk about area:

flat **space** **large** **covers more than** **small** **covers less than** **covers the same as**

the area of a park

the area of a wall

the area of a rug

the area of a coin

COMPARING 2 AREAS

It is easy to compare 2 areas.

Are the areas the same size, smaller or larger than each other?

Estimate first, then see if you can put one on top of the other to check.

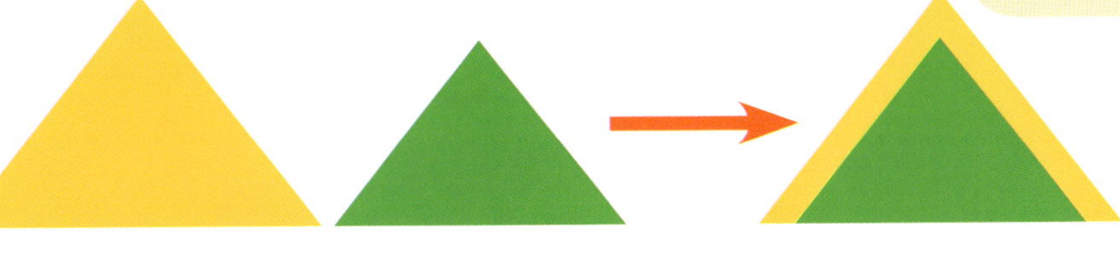

The yellow shape has the larger area.

Try to match edges together.

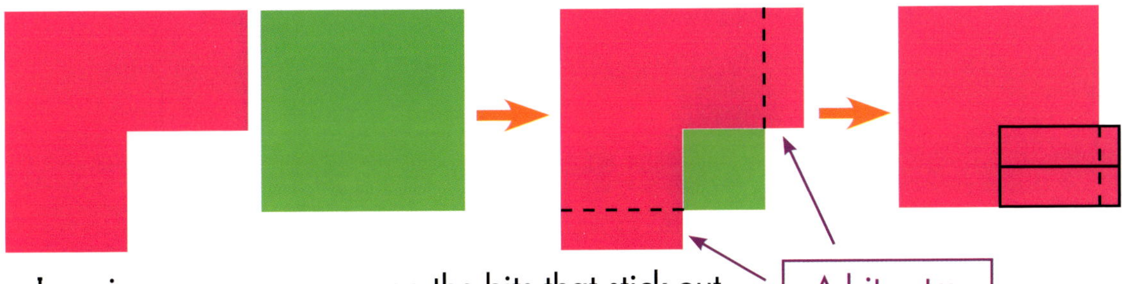

Imagine you can rearrange the bits that stick out.

The green area is a little bit smaller than the pink area.

Sometimes you can cut up one area and put the pieces on top of the other area.

These 2 areas are about the same.

Try this

Find 2 different objects that you estimate have the same area.

Explain how to find out if one area is larger, the same size as or smaller than the other.

EXPLORING TESSELLATION

You can cover an area with any flat shape.

This beautiful wall is covered with broken tiles.

You **tessellate** when you cover an area with shapes so there are no gaps and no overlaps.

These tiles don't tessellate.

There are no overlaps, but look at the gaps left in between.

Circles don't tessellate.

They leave too many gaps.

Identical squares tessellate.

They are a favourite shape for tiling bathrooms and paths.

Regular hexagons tessellate.

Identical oblongs tessellate.

MEASURING AREA WITH UNITS

You can use the area of small objects to measure and compare larger areas.

Try to use identical shapes.

Some shapes work better than others.

Shapes that tessellate work best

Harry covered his rug with bones.

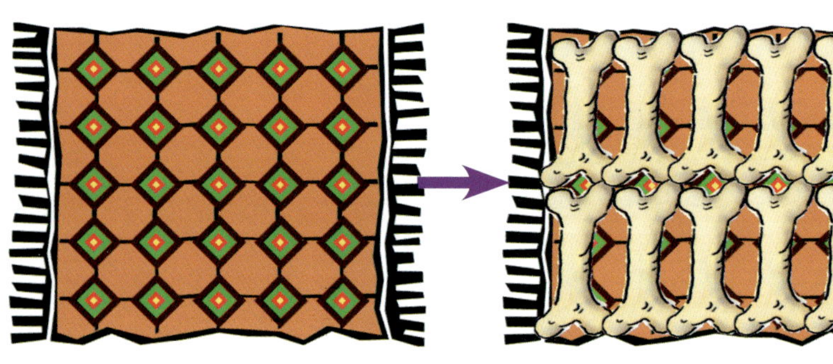

Harry's rug has an area of 10 bones.

Harry covered the floor of his kennel with bones.

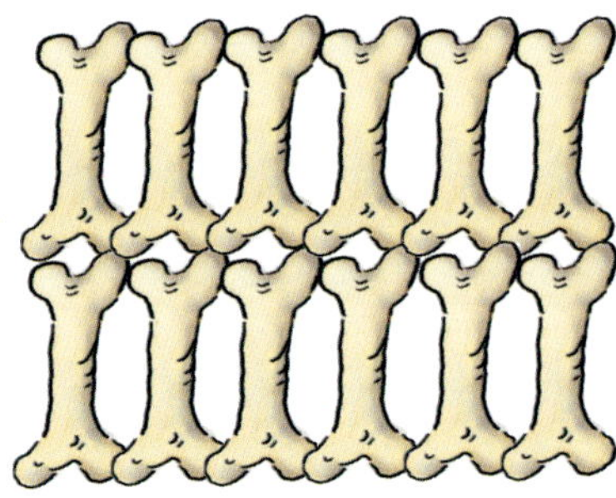

It covers an area of 12 bones.

The floor has a larger area than the rug.

Naja estimates her handprint covers 10 square units.

She doesn't have enough units to check her estimate.

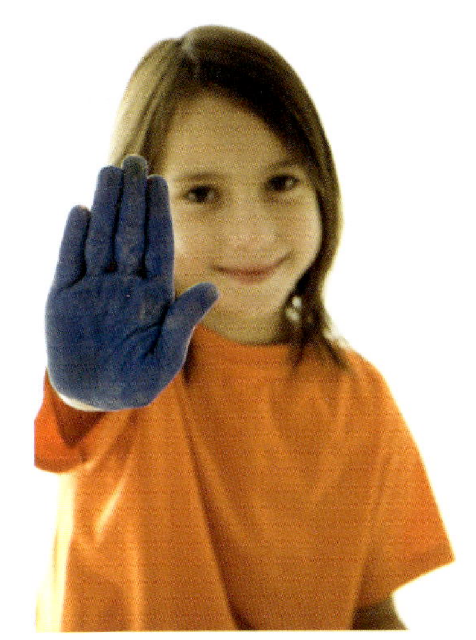

She needs about 13 units to cover her handprint.

The larger area needs more units to cover it

You can make your own square area units.

Find a shape you can trace onto paper.

a square base

a square base

Try this

How many book units cover your desk?

Is this area smaller than, the same as or larger than your body area?

Estimate first, then check.

Explain your strategy to a friend.

MEASURING AREA WITH A GRID

Instead of covering flat shapes with same-size units, you can use a **grid** to measure area.

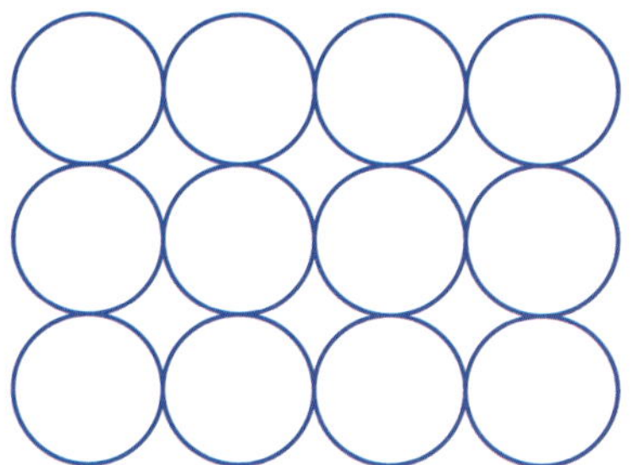

a grid of circles

a grid of triangles

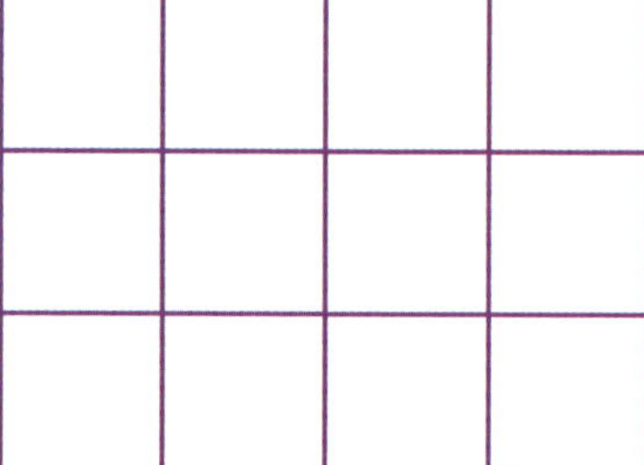

a grid of squares

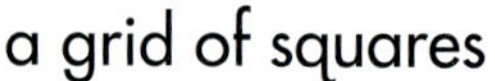

10 areas the size of Turkey cover the same area as Australia.

Lay an object on top of your grid.

Trace around it with a pencil.

Remove your object.

Count the number of grid units covered.

Remember that bits left over might be half or quarter units.

Larger areas cover a larger number of units on the grid.

Or cover your flat object with a transparent grid and count how many units of area cover your shape.

This hat is covered with a square grid.

If you could turn it around you could count all the square units of area.

This $5 note covers about 10 triangles and 4 half-triangles.

It has an area of about 12 triangles.

Nadia's beach towel covers the same area as 3 rows of 8 Maths Guides. That's 24 units altogether.

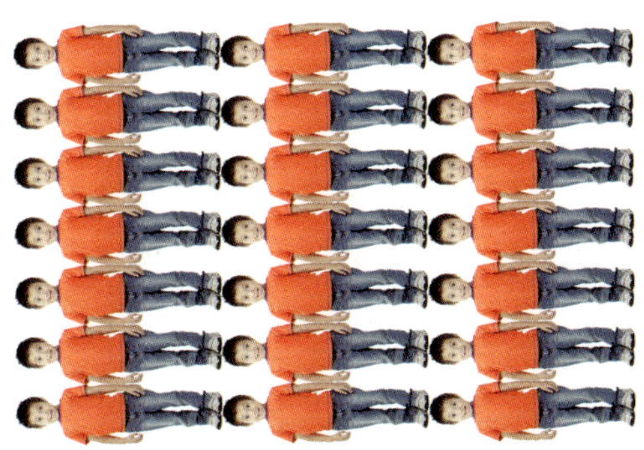

The area of Jack's bedroom is the same as 7 rows of 3 Jacks. That's 21 Jacks altogether.

Challenge

Look at the square grid covering every face of this cube.

How many square units in the total outside area of the cube?

____

Mass is a measure of how heavy an object feels

MASS

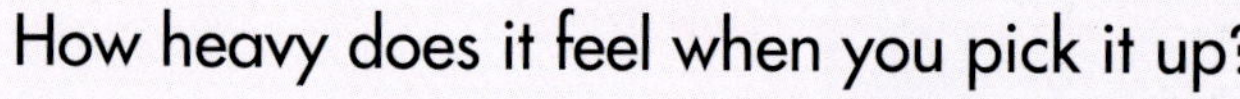

How heavy does it feel when you pick it up?

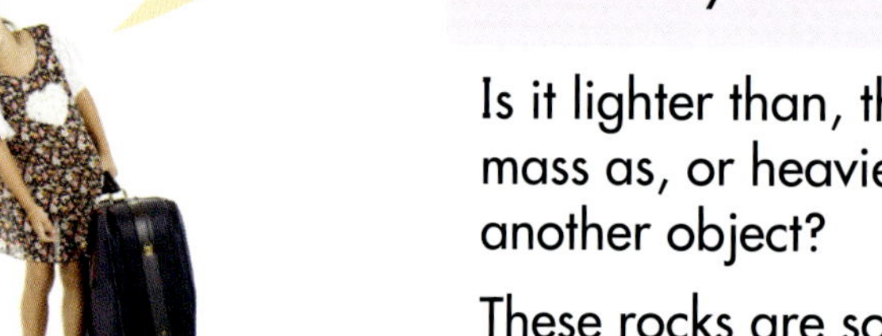

Is it lighter than, the same mass as, or heavier than another object?

These rocks are so heavy they need a truck to lift them.

These books feel as heavy as the box.

A blue whale is the heaviest animal on Earth.

This pumpkin feels lighter than this pineapple.

These words help you talk about mass:

lift **pull** **large** **heavy** **balance** **light** **find the mass of** **heft**

Mass is not a measure of how large something is.

This small sphere is heavier than this large ball.

Mass is not a measure of how many things you have.

This pumpkin is heavier than these 7 bananas.

COMPARING MASS

Put an object in both hands.

Compare how heavy they feel.

It helps if you close your eyes too.

This sphere feels lighter than this cube

Some differences in mass are obvious.

This elephant is heavier than this rock.

You don't need to pick them up to compare their mass.

But if the difference is close, you measure to know which one is heavier.

Joe can pick up the boxes.

Joe can't pick up the rock.

The boxes are not as heavy as the rock.

Which bag do you think looks heavier?

You can't tell until you pick them up.

The orange bag is heavier.

Find 3 objects you think have a different mass.

Estimate their order from the lightest to the heaviest.

Heft to check.

Try this

It is not easy to feel the difference in mass if you heft light objects

USING A MASS BALANCE PAN

A balance device helps you compare mass.

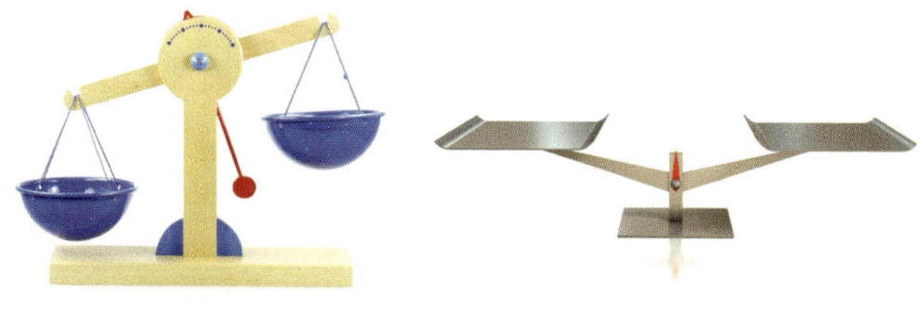

an equal arm balance

a rocker balance

Put an object in each side of the balance.

Watch what happens.

A heavier object makes one side go down.

Objects with the same mass make both sides stay the same.

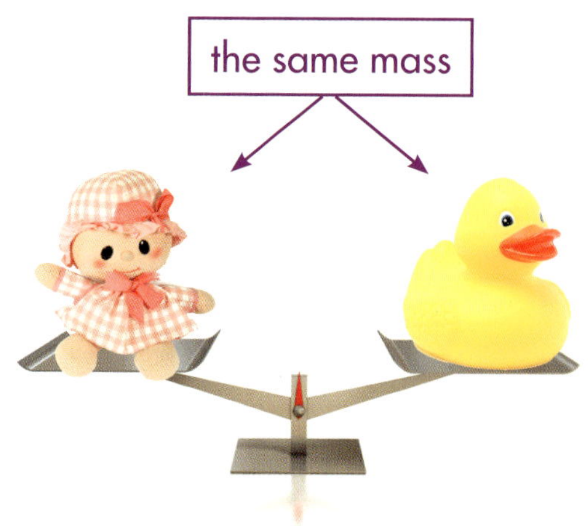

MEASURING MASS WITH UNITS

Try to find objects that look the same to use as units.

bottle tops nuts shells balls pebbles toy cars

Estimate how many mass units balance an object.

You can compare mass using smaller mass units

Ali thinks this toy is as heavy as 8 shells.

He checks using a balance.

He puts the toy on one side.

He counts as he put shells on the other side.

Ali uses 10 shells to balance this toy.

Lucy used 8 shells to balance this toy.

It is lighter than Ali's toy.

Joey measures his mass on a bathroom scale.

It measures in units called kilograms.

Joey's mass is 15 kilograms.

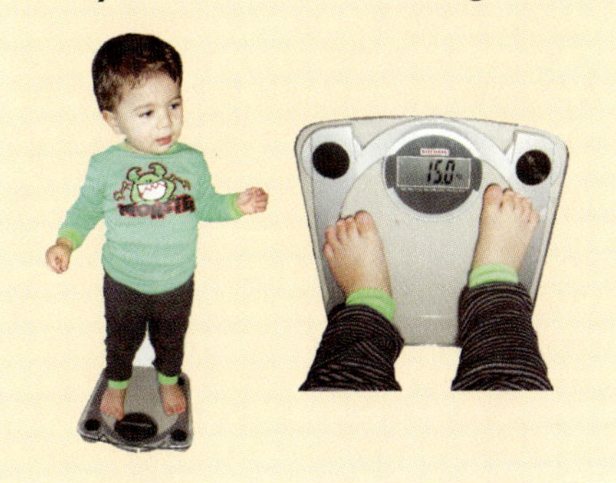

Select a mass unit.

Estimate the number of units to balance 3 different objects.

Check with a mass balance pan.

The heaviest pumpkin is about 825 kilograms.

PUT MASSES IN ORDER

Measure how many mass units you need to balance different objects.

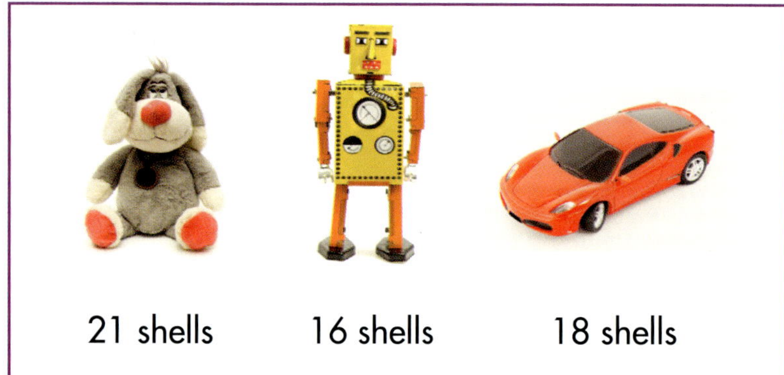

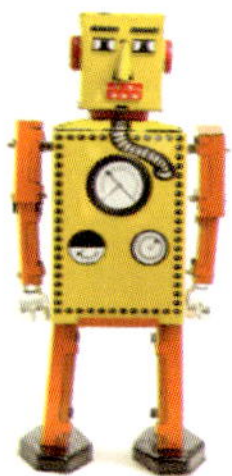

You can now put them in order from the lightest to the heaviest.

The object with the least units is the lightest.

The robot is the lightest.

The object with the most units is the heaviest.

The rabbit is the heaviest.

Some objects have exactly the same number of units.

This zebra has the same mass as the robot.

It has a mass of 16 shells.

Estimate the mass of the toy dog in shells.

It is heavier than the car.

It is lighter than the rabbit.

____

VOLUME AND CAPACITY

How large is it?

It does not measure heaviness.

Large volumes

Capacity is a measure of the volume inside a container.

How much can a container hold?

Large capacities

Here are some words to help you talk about volume and capacity.

full

half-full

large

is smaller than

amount

empty

overflowing

holds more than

FULL, HALF FULL AND EMPTY

The **rim** is the top of a container.

Some things can stack or pack above the rim when they are full.

Liquids like water or juice are under the rim when they are full, otherwise they will spill.

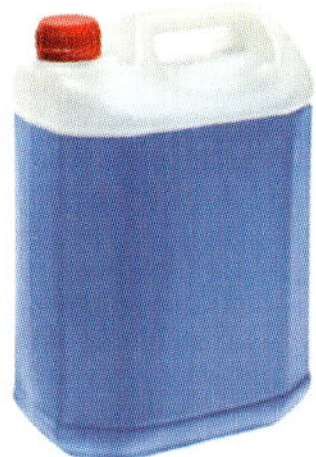

Some things pack exactly to the rim with nothing overflowing.

Two equal parts are 2 halves.

These containers are half full.

The containers below are empty.

COMPARE VOLUME

Look at 2 different objects.

Estimate their size.

Which one is larger?

Or are they about the same size?

Some volumes are obviously larger than others.

Some volumes look about the same size.

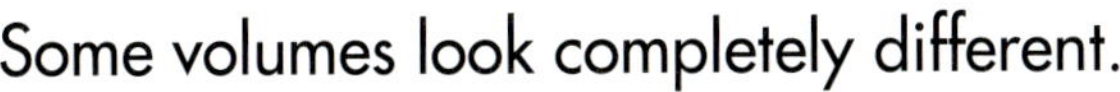

Some volumes look completely different.

Can you imagine one shape fitting inside the other shape?

If so, it has a smaller volume.

Can you imagine rearranging different parts to make the shapes look similar?

If so, the volumes are about the same.

What is the largest bubble you can blow?

Find 3 objects that you estimate have a volume similar to this bubble.

COMPARE CAPACITY

You put objects in order by how much they can hold.

Even if a container is empty or only a little bit full, the capacity is the total amount it can hold when it is full.

Some capacities are obviously larger than others.

Some capacities look about the same size.

Look at these 2 jugs.

Estimate which one can hold more.

Or can they hold about the same amount?

The juice jug is not full yet.

It has the capacity to hold more juice.

The milk jug is pretty full.

How can you measure the capacity of each jug?

Here is one way:

When they are empty, fill them up with water.

Pour the water from the first jug into a big bowl.

Look where it reaches then pour it out.

Pour the water from the second jug into the bowl.

Look where it reaches.

Is it higher than the first jug's mark?

If so, this jug holds more.

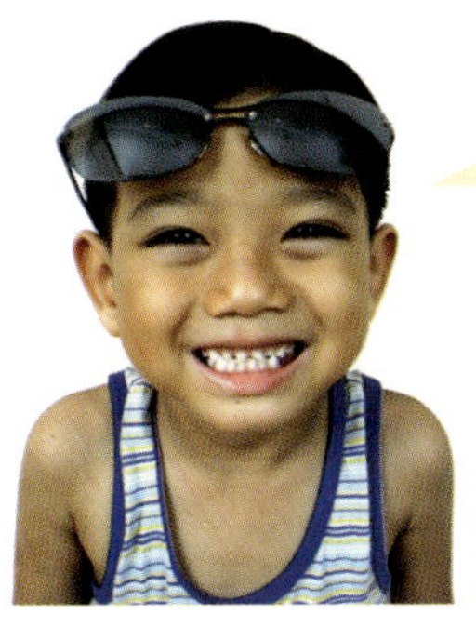

I used this strategy to compare the capacities. The milk jug holds a little bit more.

So the milk jug has the larger capacity

Try this

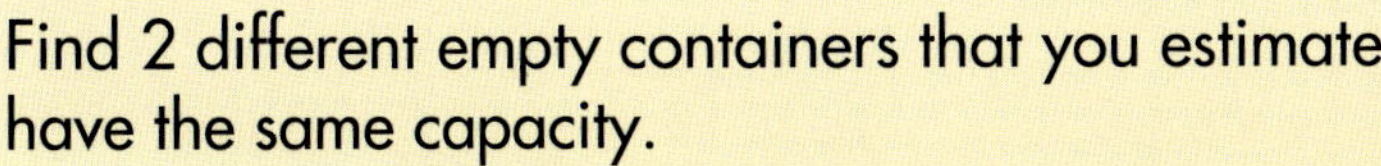

Find 2 different empty containers that you estimate have the same capacity.

Check which holds more water.

Try units that stack together

MEASURE VOLUME WITH UNITS

You can compare volume using smaller volume units.

Try not to leave any gaps.

Blocks stack easily without gaps.

Try to use same size blocks.

Estimate how many blocks you need before you start to build.

Then stack blocks to match the volume.

This stack uses 6 bricks.

Smaller units use more blocks

This stack uses smaller units.

It uses 27 small cubes.

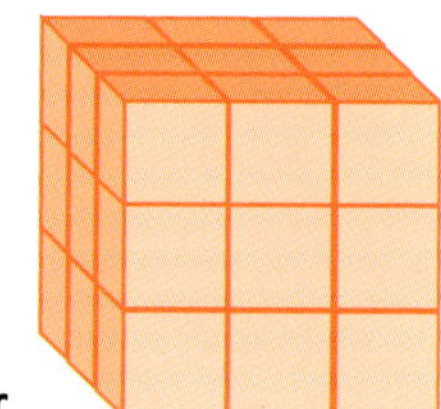

Both cube stacks have the same volume, even though the number of units is different.

Larger units use fewer blocks.

This ice-cream stack uses 12 bricks.

It is double the volume of the cube above.

Try this

Find 2 different objects that you estimate have the same volume.

Make models using your own volume units to check.

MEASURE CAPACITY WITH UNITS

You can compare capacity using smaller capacity units.

Here are some common units of capacity:

This egg cup holds 10 spoons of water.

This jar holds 12 cups of flour.

This pot holds 14 bowls of soup.

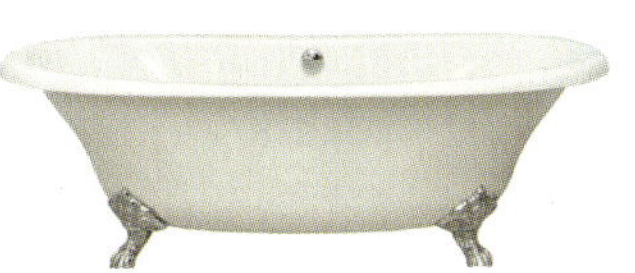

This bath holds 17 buckets of water.

Try this

Make your own capacity measuring device.

Take an empty plastic bottle. Mark the height of 1 cup of water.

Continue adding 1 cup at a time until it is full.

Use this to measure and compare other capacities in cups.

Challenge

Make a list of your favourite recipes that use cups and spoons.

Banana Muffin

2 cups of self-raising flour
1 cup of milk
½ cup of sugar
¼ cup of melted butter
2 beaten eggs
1 cup of mashed banana
½ teaspoon of cinnamon spice

Turn oven to 175 degrees.
Brush muffin tray with melted butter.
Mix flour and sugar in a bowl.
Stir in milk, melted butter and beaten eggs.
Add mashed banana. Add cinnamon.
Spoon into muffin tray.
Bake 15 minutes.

Makes 12 Muffins

TELL THE TIME

People love to measure time passing.

You measure this in hours and minutes, days, weeks, months and years.

These words help you talk about time:

late, on time, early, yesterday, o'clock, half past, today, tomorrow, a quarter to, a quarter past

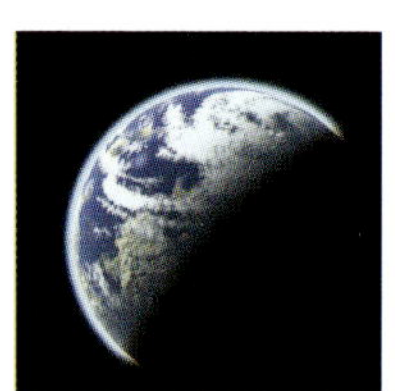

It takes 24 hours for the Earth to rotate once.

clockwise

READ TIME ON THE HOUR

An analogue clock has a round face divided into 12 equal parts.

Each part represents 1 hour passing.

Analogue clocks have 2 rotating hands.

The direction the hands rotate is called **clockwise**.

You tell the time by the position of the hands.

On the hour, the longer hand points to 12.

The shorter hand points to the hour number.

Digital clocks measure hours using numbers only.

The hour numbers are on the left. **01** to **12:**

The minute numbers are on the right. **:00**

You say: ▲ **o'clock**

You write ▲ **:00**

'Clock' comes from an old Dutch word for bell.

Midnight is **12 o'clock**.

It is the middle of the night.

It is **12:00** midnight.

Rex's alarm is set for **7 o'clock**.

He'll wake up at **7:00** in the morning.

Midday is **12 o'clock** too.

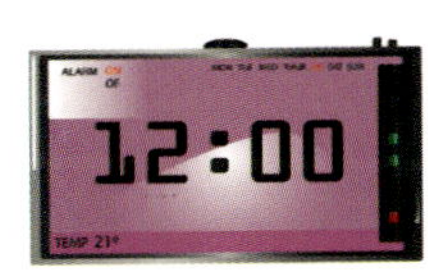

Long ago people measured hours by burning a candle.

Dina and Ben eat lunch in the middle of the day.

It is **12:00** midday.

Mei reads her book from **2 o'clock** to **3 o'clock**.

She reads for 1 hour.

Her clock doesn't have any numbers.

She just looks at the position of each hand.

Omar watches TV from **5 o'clock** to **7 o'clock**.

That's 2 hours from **5:00** to **7:00**.

Write the digital time under each clock.

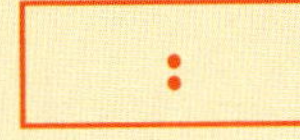

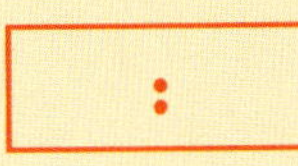

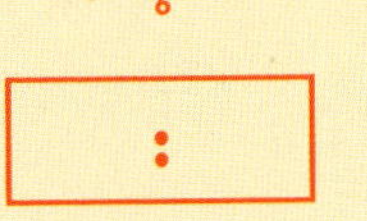

Try this

HALF PAST

An hour can be divided into 2 equal parts.

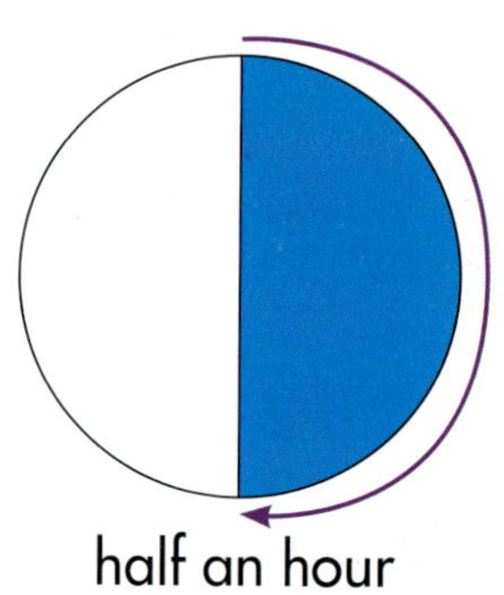

Half past means half an hour past the last hour.

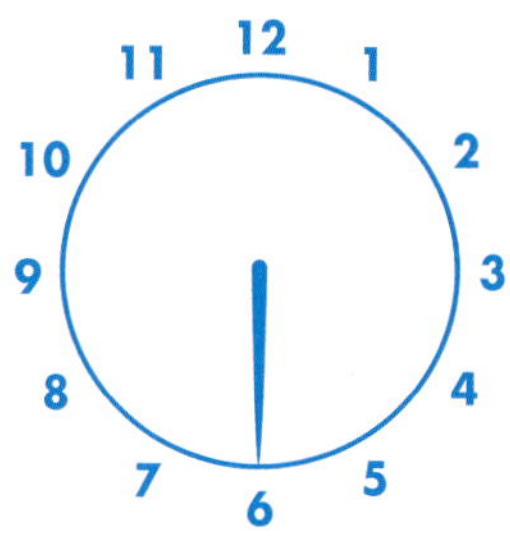

The longer hand points to 6.

The shorter hand is halfway between the last and the next hour.

There are 60 minutes in 1 hour.

30 + 30 minutes

You say: **half past ▲** or **30 minutes past ▲**

You write: **▲ :30**

Lara takes half an hour to make and bake muffins.

She starts at **10 o'clock**.

She finishes cooking at **10:30**.

That's **thirty minutes past 10**.

Lucy and Pat shop from 12 midday to **12:30**.

They take half an hour.

It's now **thirty minutes past 12**.

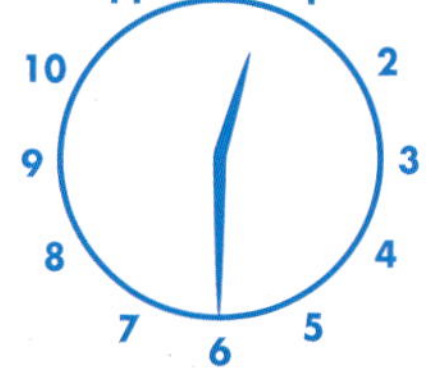

Gabi cleans her teeth at **8 o'clock**. She goes to sleep half an hour later.

That's ☐ **minutes past** ☐

That's ☐ **:30** at night. ☐ : ☐

A QUARTER PAST

An hour can be divided into 4 equal parts.

There are 60 minutes in 1 hour.

15 + 15 + 15 + 15 minutes

A quarter of an hour = 15 minutes

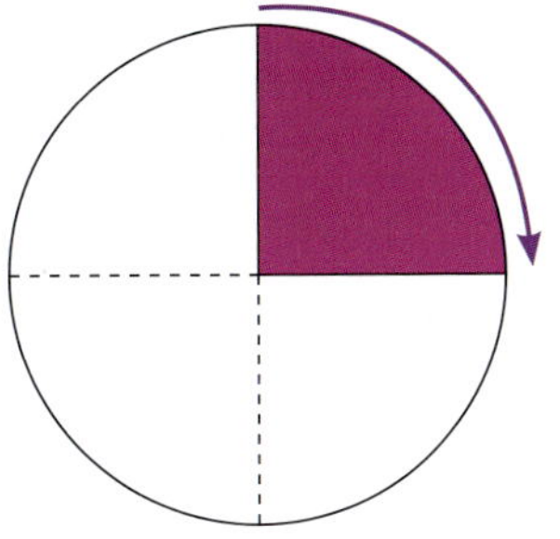

a quarter of an hour

A quarter past means 15 minutes past the hour.

The longer hand always points to 3.

The shorter hand is just past the hour.

You say: **a quarter past** 

or **15 minutes past** 

You write: ▲ **:15**

Lee builds with blocks from 9:00 until **9:15**.

That's 15 minutes.

She finishes at **a quarter past 9**.

Sara kicks a ball for 15 minutes.

She starts at **1 o'clock**.

She finishes at **15 minutes past 1**.

That's **1:15** in the afternoon.

Oscar takes 15 minutes to eat his yoghurt.

He starts at **5 o'clock**.

He finishes at **a quarter past** 

That's ☐ **:15.**

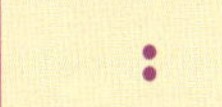

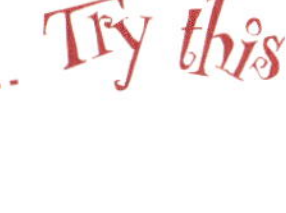

A QUARTER TO

A quarter to means 15 minutes before the next hour.

a quarter to

That's the same as 3 quarters of an hour past the last hour.

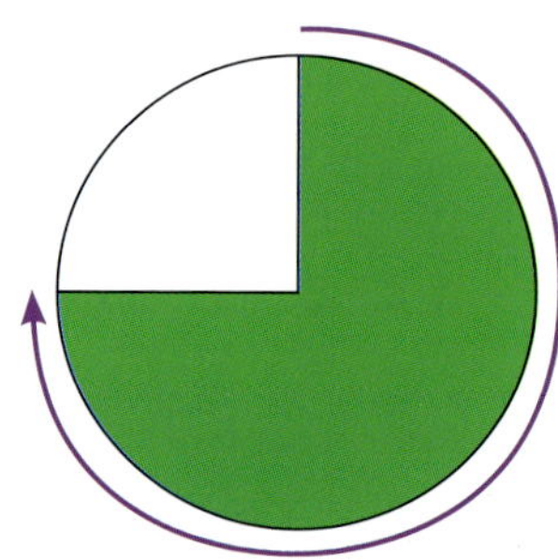

3 quarters of an hour

15 + 15 + 15 = 3 quarters

'45 minutes past' the last hour is the same time as 'a quarter to' the next hour

The longer hand always points to 9.

The shorter hand is just before the next hour.

You say: **a quarter to ▲**

or **45 minutes past ▲**

or **▲ forty-five**

You write: **▲ :45**

Joey always has his dinner at **a quarter to 6**.

That's **45 minutes past 5**

That's **five forty-five**

That's **5:45**

Leo walks his dog from 11:00 to **11:45**.

That's 45 minutes.

He walks until **a quarter to 12**.

Harry practises his dance for 45 minutes.

He starts at 3 o'clock and finishes at **3:45**.

Harry stops at **a quarter to 4**.

The vet checks Abba from 5:00 to ☐ **:45**.

05:45

That's 45 minutes.

She finishes at ____ ____ ____ **6**.

Try this

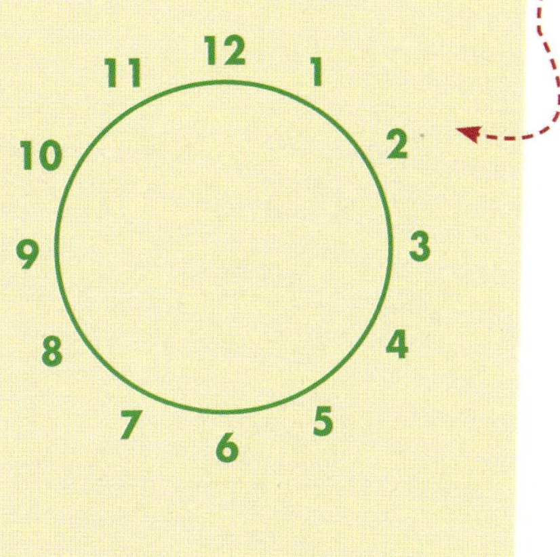

The moon takes about 4 weeks to circle planet Earth.

READ A CALENDAR

DAYS OF THE WEEK

Each week is divided into 7 equal parts called **days**.

Imagine the days in a circle.

The days repeat like this forever.

Monday to Friday are **week days**.

Saturday and Sunday are the **weekend**.

What does the name of each day of the week mean?

Wednesday is named after the old Norse god Wodin.

Challenge

This sign shows when Sally's shop is open.

Tick the times below when you can buy flowers.

☐	Mon	9:00
☐	Tues	11:30
☐	Wed	3:00
☐	Thurs	10:45
☐	Fri	9:15
☐	Sat	12:45
☐	Sun	4:00

OPENING HOURS:

Mon.	10.00	to	4.00
Tues.	9.00	to	2.00
Wed.	9.00	to	2.00
Thur.	9.00	to	9.00
Fri.	8.30	to	2.00
Sat.	8.30	to	2.00
Sun.	Closed		☹

MONTHS AND SEASONS

The **year** is based on the rotation of planet Earth around the Sun.

Each year is divided into 12 parts called **months**.

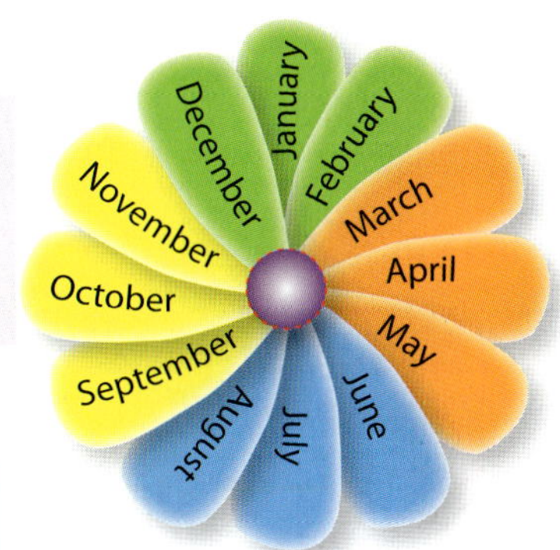

The number of days in a month varies.

	Name of month	Number of days	Season
1	January	31	Summer
2	February	28	Summer
3	March	31	Autumn
4	April	30	Autumn
5	May	31	Autumn
6	June	30	Winter
7	July	31	Winter
8	August	31	Winter
9	September	30	Spring
10	October	31	Spring
11	November	30	Spring
12	December	31	Summer

Planet Earth takes 12 months to travel around the sun.

August is named after the old Roman emperor Julius Caesar

There are 365 days in 12 months.

Every 4th year is a leap year.

You add an extra day to February.

February has 29 days in a leap year.

There are 366 days in a leap year.

It takes about 1 year for a baby to learn to walk.

The Summer Olympic Games are held in a leap year.

FOUR SEASONS

Each year is divided into 4 seasons.

Spring	Summer	Autumn	Winter

These seasons are based on how far or close Earth is to the sun as it rotates.

In Australia :

- **Summer months are usually hot.**
- **You can go to the beach.**

- **The autumn months are cooler.**
- **It can be windy too.**

- **Winter is the coldest season.**
- **You wear warm clothes.**

- **Spring is warmer.**
- **Lots of flowers grow in spring.**

READ A CALENDAR

A **calendar** shows the days of the year in order.

It lists the months from January to December.

The rows show the days in each week.

The columns show the days with the same name.

The **date** is the day, month and year for each day.

You read each number as a position word.

3 is 'the third', 19 is 'the nineteenth' and 24 is 'the twenty-fourth'.

1 January is the first day of each year.
It is called New Year's Day.

Omar's 7th birthday is Wednesday the sixteenth of May 2012.

31 December is the last day of the year.

It is called New Year's Eve.

People all over the world celebrate the start of a new year just after midnight.

The Chinese calendar is based on the cycle of both the moon and the sun.

Chinese New Year is later in January or February.

When is Chinese New Year this year?

The top row in each month shows the name of the day

READ A CALENDAR (continued)

JULY						2013
SUN	MON	TUE	WED	THU	FRI	SAT
	1	2	3	4	5	6
7	8	9	10	11	12	13
14	15	16	17	18	19	20
21	22	23	24	25	26	27
28	29	30	31			

The first day of July 2013 is a Monday.

The day before belongs to June.

11 July is the second Thursday.

Wednesday 31 July is the last day of the month.

The last 3 days in the 5th week belong to August.

Look at the July 2013 calendar.

Jake gets his bike serviced on the 4th Friday. Circle this date.

Great Gran's 80th Birthday is the third last day of the month.

Cross out this date.

Challenge

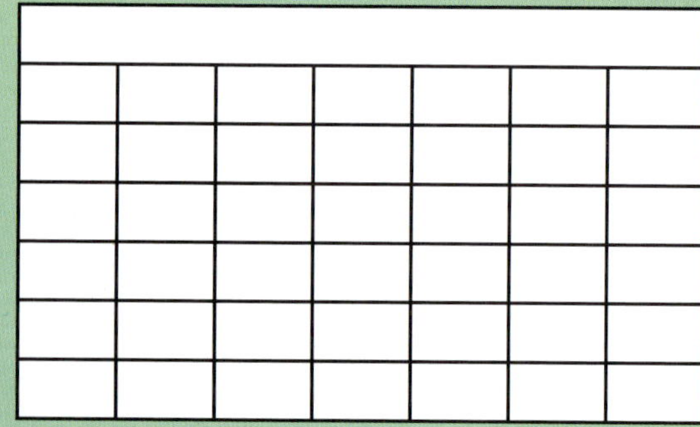

Create your own calendar for your birthday month.

Fill in the days of the week.

What day is the 1st of your month?

Write 3 questions to ask someone about your calendar.

3D OBJECTS

Let's find out about the shapes of things

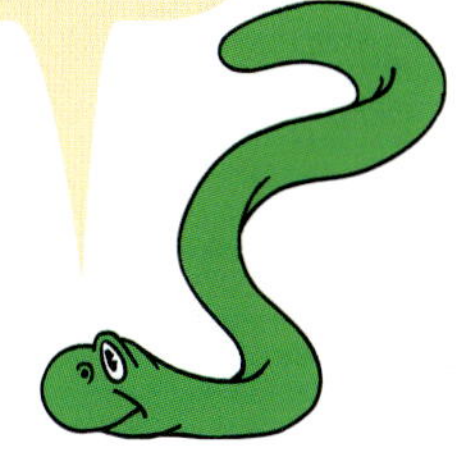

SPHERES

Balls look like spheres.

A sphere:

- can roll
- is completely round

3D objects are **spherical** if they look like a sphere.

- can be any size.

A sphere has:

- no flat **faces**
- no **corners**
- no straight **edges**
- 1 curved **surface**.

curved

When you draw a sphere it looks like a circle.

You can make a pattern with spheres.

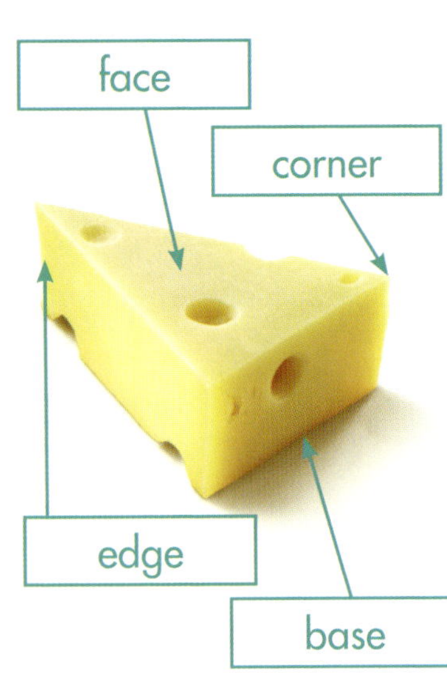

PRISMS

A **prism** has 2 flat bases joined by 2 or more rectangular faces.

The name of a prism comes from the shape of the base.
The 2 bases are opposite each other.

A prism is not a pyramid.

Common prisms are:

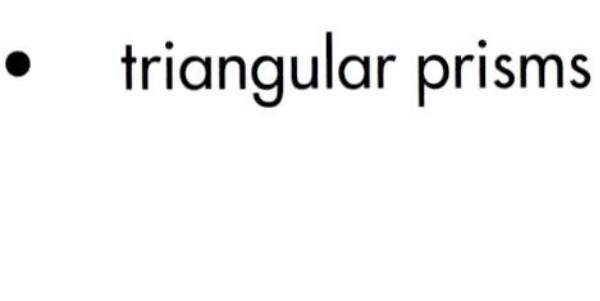

- triangular prisms

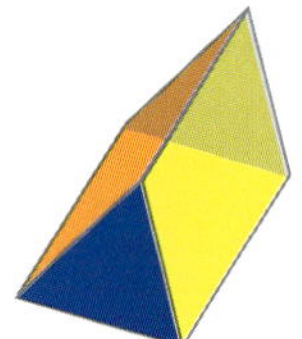

- rectangular prisms

- pentagonal prisms

- hexagonal prisms

- octagonal prisms

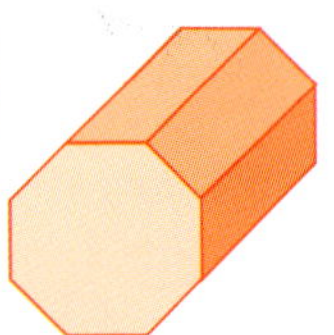

face: a flat 2D surface on a 3D object

corner: the point where 2 or more straight lines meet

edge: the line on a 3D object where 2 faces meet

This view of the cube shows 2 faces, 6 corners and 7 edges.

But this cube has:

- 4 more hidden faces
- 2 more hidden corners
- 5 more hidden edges.

Imagine you can see them all.

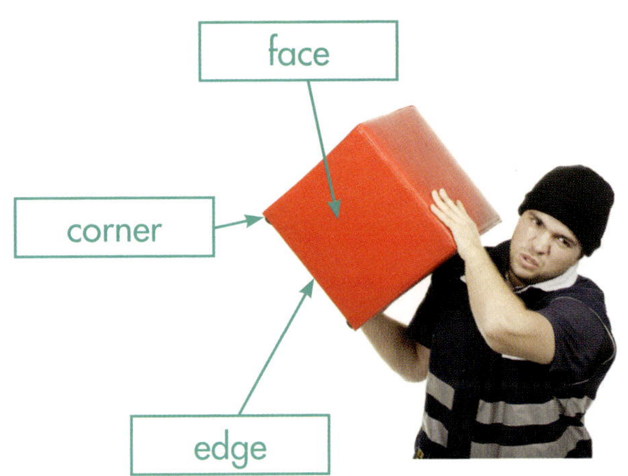

Count all the hidden bits too

Name of prism	Shape of base	Number of faces	Number of corners	Number of edges
triangular		5	6	9
rectangular		6	8	12
pentagonal		7	10	15
hexagonal		8	12	18
octagonal		10	16	24

CUBES

A **cube** is a special rectangular **prism**.

Every **face** is the same shape and size.

Every **corner** looks the same.

Every **edge** is the same length.

It has square bases.

Cubes are 3D objects

This box looks like a cube.

Cubes:

- can't roll but can stack

- can be any size

- can be built from smaller blocks.

A cube has:

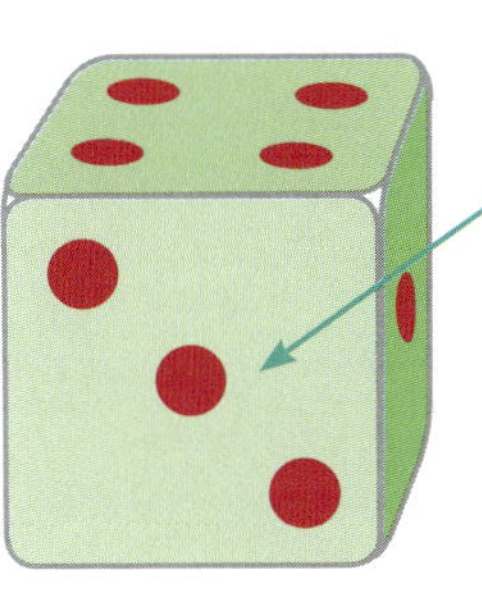

- 6 square faces
- 8 corners
- 12 straight edges.

You can make a model of a cube from sticks and plasticine:

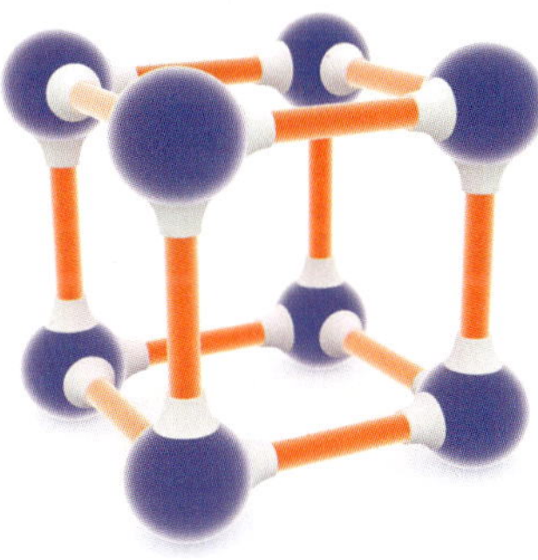

You can draw a cube in different positions:

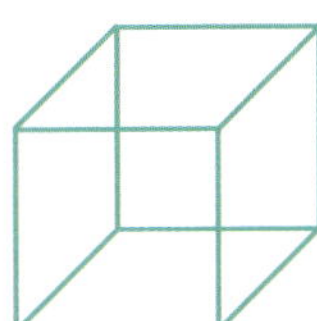

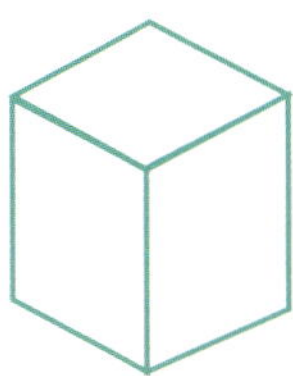

You can make a pattern with cubes.

3D objects are **cubical** if they look like a cube.

CYLINDERS

A **cylinder** is a special 3D object.

It is like a circular **prism**.

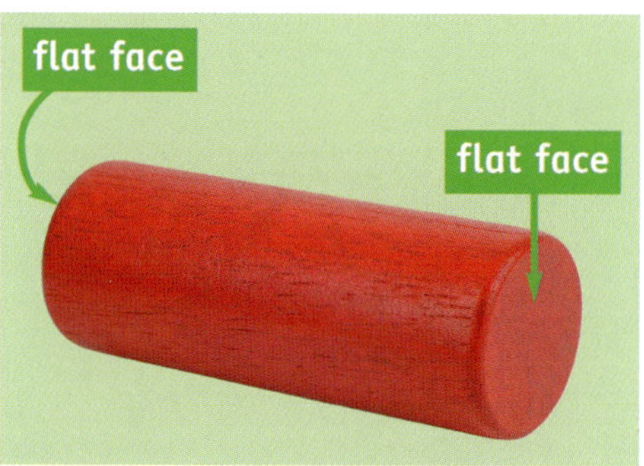

It has 2 flat bases opposite each other.
These are joined by 1 curved surface.

This can looks like a cylinder.

Cylinders:

- can roll and stack

- can be any size or thickness.

A cylinder has:

- 1 curved surface
- 2 circular faces or bases
- 0 corners
- 0 straight edges.

You can draw a cylinder in different positions:

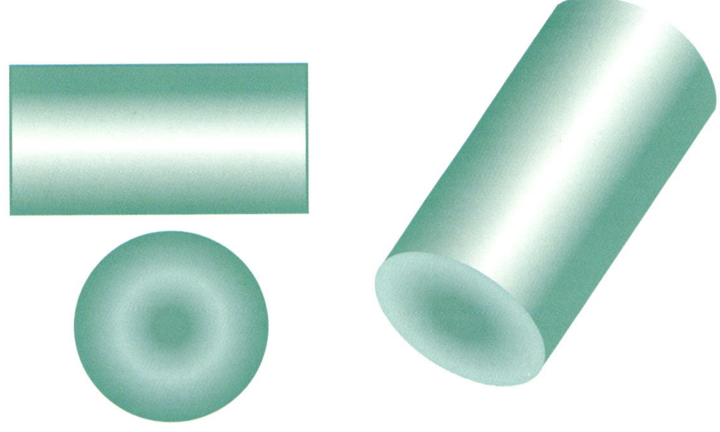

3D objects are **cylindrical** if they look like a cylinder.

You can make a pattern with cylinders.

PYRAMIDS

A **pyramid** has 1 flat base joined by 3 or more triangular faces that meet at a point.

A square pyramid is the most common pyramid.

This tower looks like a square pyramid.

A pyramid is not a prism.

The most famous square pyramids in the world are in Egypt.

The name of a pyramid comes from the shape of the base.

Common pyramids are:

- triangular pyramids

- rectangular pyramids

- pentagonal pyramids
- hexagonal pyramids
- octagonal pyramids.

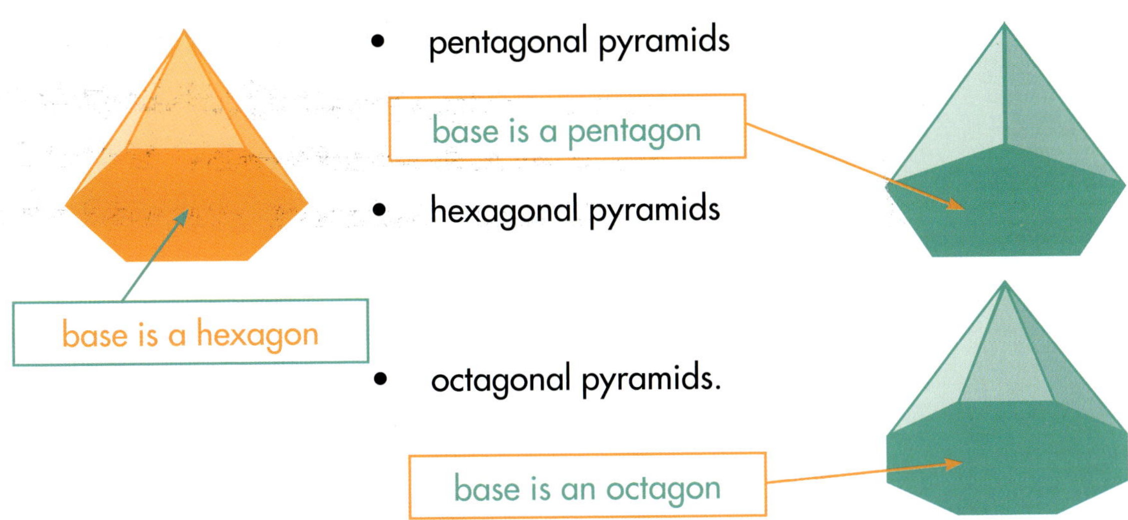

Name of pyramid	Shape of base	Number of faces	Number of corners	Number of edges
triangular		4	4	6
square		5	5	8
rectangular		5	5	8
pentagonal		6	6	10
hexagonal		7	7	12
octagonal		9	9	16

CONES

A **cone** is a special 3D object.

It is like a circular **pyramid**.

It has 1 base but with 1 curved surface that ends at a point.

The pointy top of a cone or a pyramid is an **apex**.

An umbrella can look like a cone.

Cones:

- can roll

- can be any size or thickness.

3D objects are **conical** if they look like a cone.

A cone has:

- 1 circular face or base
- 1 curved surface
- 1 **apex**
- 0 straight edges.

You can draw a cone in different positions:

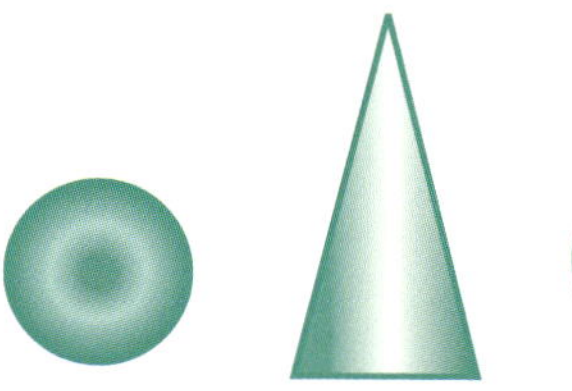

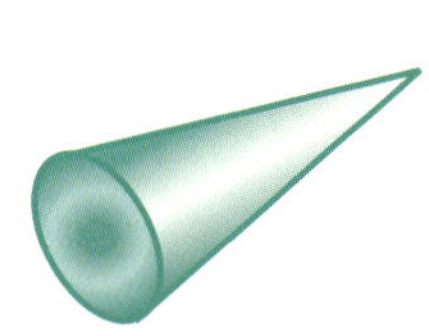

Describe each 3D object to a friend.

Don't tell them the name.

Can they guess the object?

1.  **2.**  **3.**  **4.**  **5.**  **6.** 

2D SHAPES

Two-dimensional (2D) shapes are flat.

You see them as faces on 3D objects.

You can talk about and sort 2D shapes by:

size

colour

line shape

number of sides

symmetry

Try this

Find a pile of 2D shapes. Sort them.

Describe your sorting to a friend.

Together, discuss 2 more ways to sort the same shapes.

LINES

curved lines

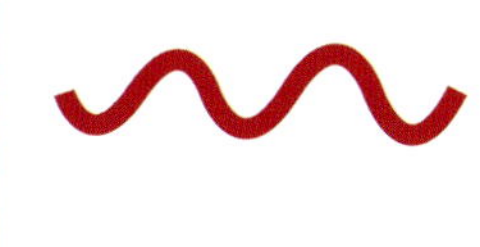

These 2D shapes all have curved edges.

straight lines

These 2D shapes all have straight sides.

STRAIGHT LINES

Straight lines can be like the horizon.

horizontal

Straight lines can be straight up and down.

vertical

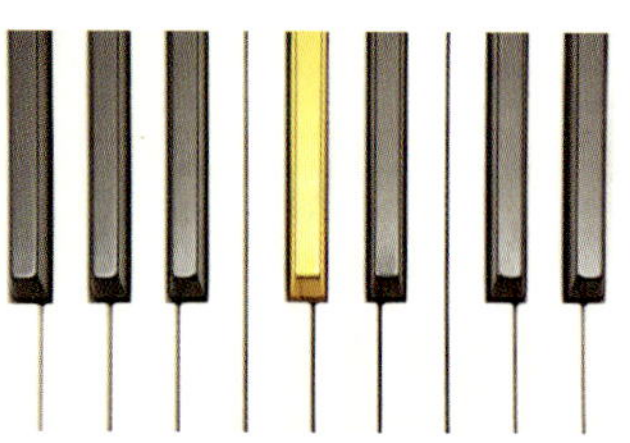

Straight lines can be leaning over.

sloping

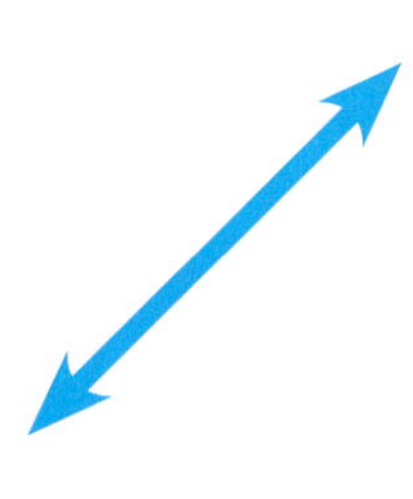

Straight lines can be the same distance apart.

parallel

The stripes on my towel are parallel

Parallel lines never touch

These straight lines are horizontal and parallel.

These straight lines are vertical and parallel.

Some of my stripes are parallel

CIRCLES

A circle has 1 curved side.

It has no corners.

You can draw a circle by tracing the base of a cylinder.

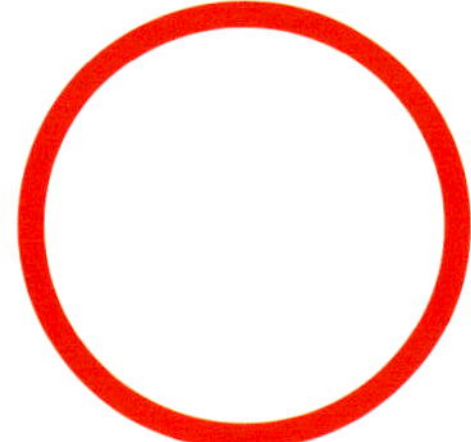

Half a circle is a **semi-circle**.

It has 1 curved edge and 1 straight side.

You can fold a paper circle into halves and quarters.

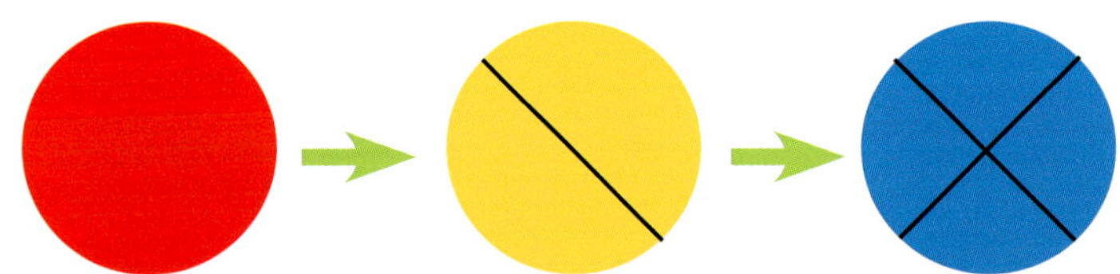

You can rearrange some shapes to create circles.

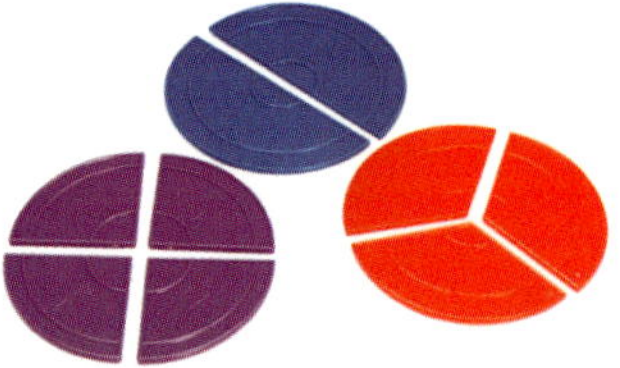

You can overlap circles to create a pattern.

STRAIGHT-SIDED SHAPES

Polygons have straight **sides**.

They have no curved sides.

Polygons have 3 or more **corners**

You count the number of sides to find the name of a polygon.

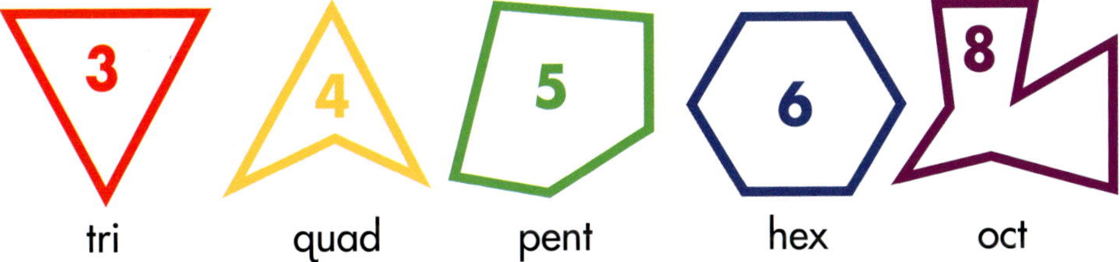

TRIANGLES

3 straight sides and 3 corners

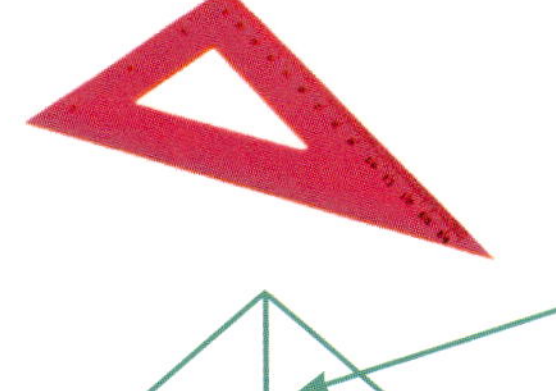

2 faces on a triangular prism

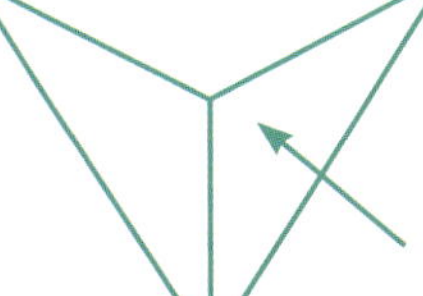

4 faces on a triangular pyramid

If you can fold a paper triangle in half it is symmetrical.

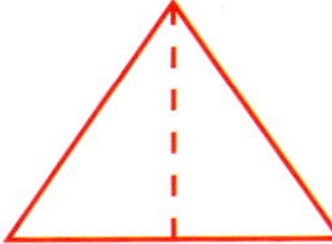

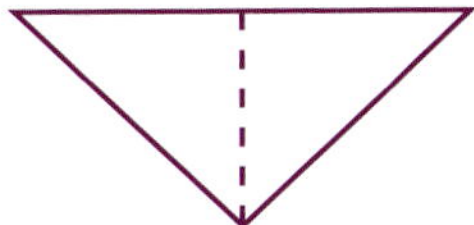

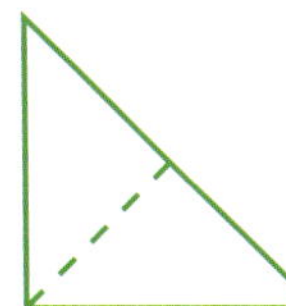

SPACE

QUADRILATERALS

4 straight sides and 4 corners

Some quadrilaterals have special names:

squares

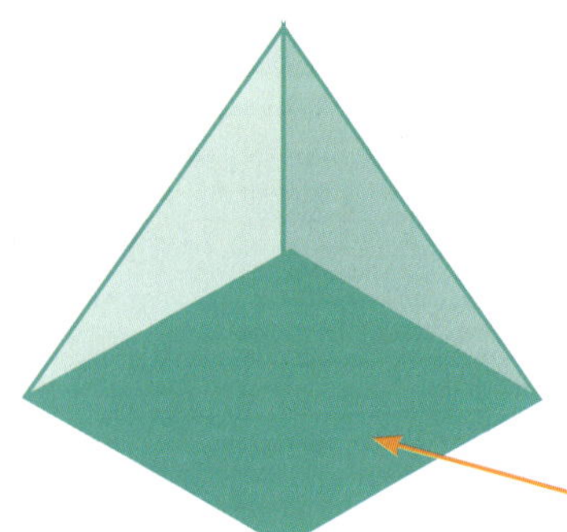

6 faces on a cube

1 face or base on a square pyramid

Every side of a square is exactly the same length.

There are 2 pairs of parallel sides.

There are 4 square corners.

Is this a square?

Measure each side to check.

No, the sides are not equal lengths.

No, this is not a square.

This is a quadrilateral.

oblongs

Squares and oblongs are both rectangles.

6 faces on a rectangular prism

6 faces on a rectangular prism

An oblong has 2 shorter sides, both the same length.

These sides are parallel.

An oblong has 2 longer sides, both the same length.

These sides are parallel.

An oblong has 4 square corners.

Is this an oblong?

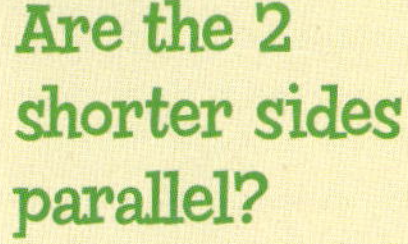

The 2 shorter sides are not parallel.

No, this is not an oblong.

This is a quadrilateral.

Make a square corner.

Fold a piece of paper in half.

Fold this in half again.

You now have a square corner.

Use this to check if a quadrilateral is a rectangle.

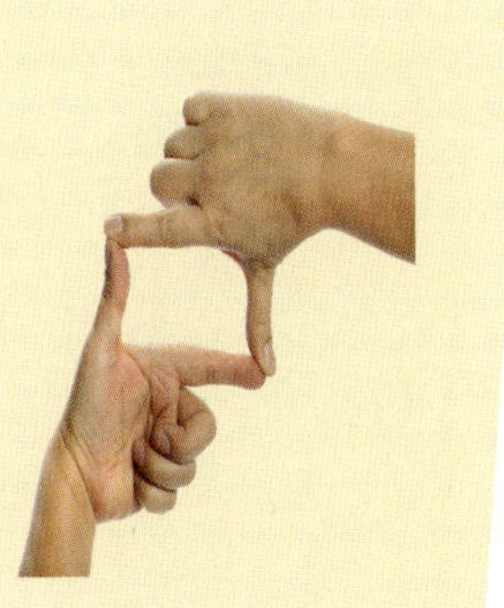

OTHER POLYGONS

pentagons

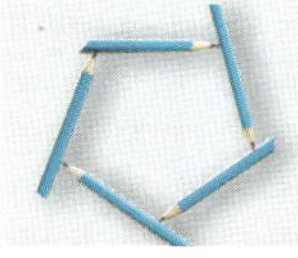   

5 straight sides and 5 corners

hexagons

   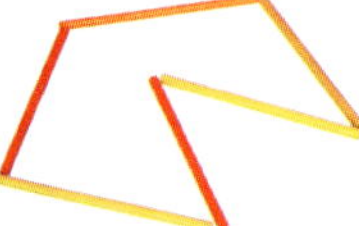

6 straight sides and 6 corners

octagons

8 straight sides and 8 corners

If you can fold a paper polygon in half it is **symmetrical**.
Some paper polygons can fold into quarters.

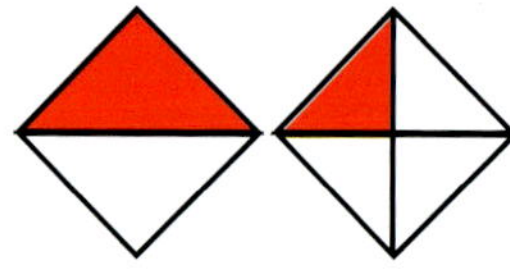

square

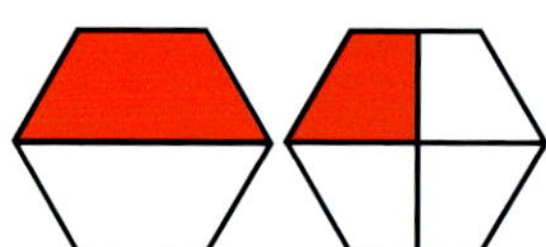

regular hexagon

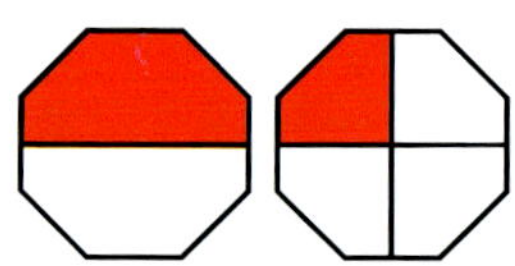

regular octagon

2D shape	Examples	Number of corners	Number of straight sides	Number of curved edges
circle		0	0	1
triangle		3	3	0
quadrilateral		4	4	0
pentagon		5	5	0
hexagon		6	6	0
octagon		8	8	0

SYMMETRY

A reflection in a mirror shows 2 halves.

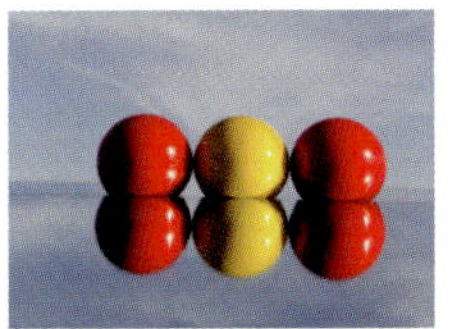

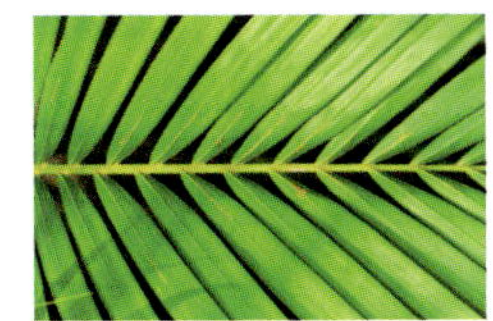

The line where 2 halves meet is the **mirror line**.

Some 2D shapes and 3D objects have an invisible mirror line. You imagine it.

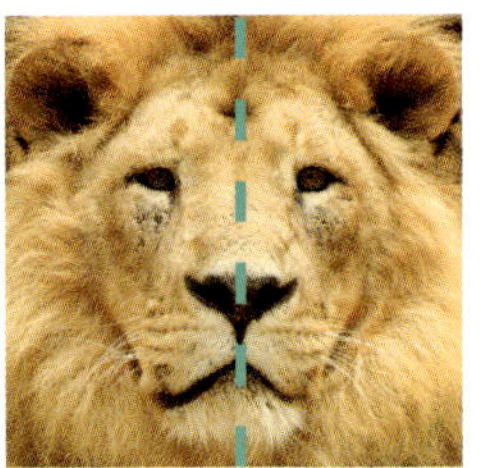

The mirror line is called a **line of symmetry**.

Some shapes have **no** line of symmetry.

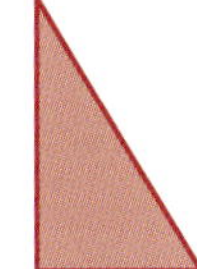

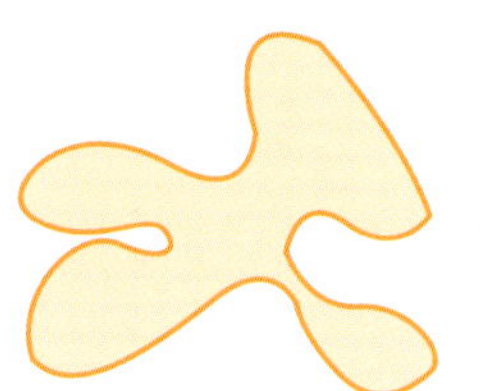

These shapes are **not** symmetrical.

A shape is **symmetrical** if 2 halves match at every point.

These are lines of symmetry.

Theses are **not** lines of symmetry.

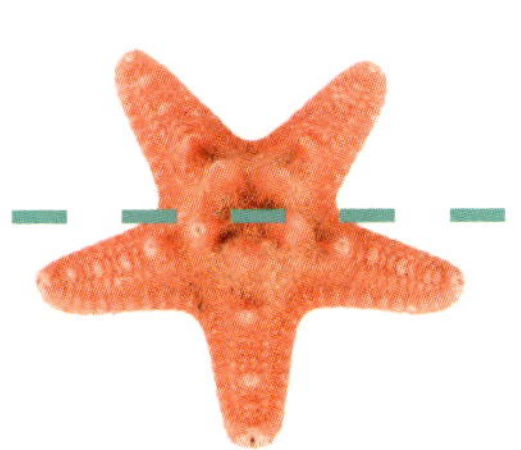

TRANSFORM 2D SHAPES

TANGRAMS

These 7 shapes are called **tangrams**.

They fit together to make a square.

You can rearrange the shapes to create your own pictures.

Rearrange up to 7 tangram pieces to create:

1. a quadrilateral

2. a pentagon

3. a hexagon

4. an octagon.

HIDDEN SHAPES

When you rearrange shapes you sometimes make hidden shapes.

Many different polygons are hiding in this biscuit shape.

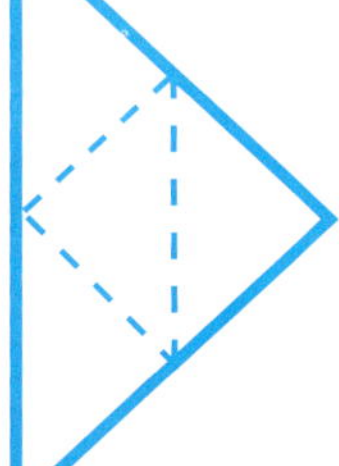

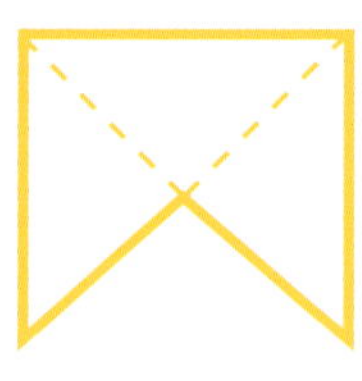

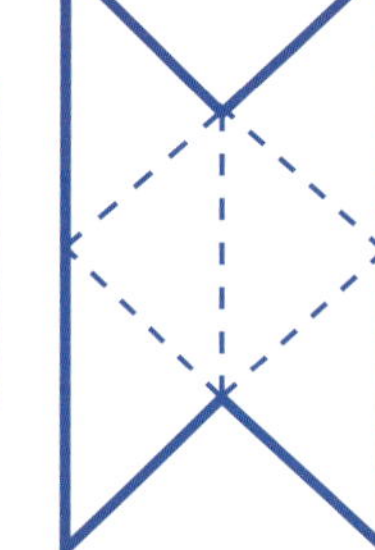

How many hidden polygons can you find in these shapes?

FLIPS

A flip pattern copies a shape across a mirror image.

You reflect a shape across a **line of symmetry**.

It can be in any direction.

You can flip any shape.

Flips create symmetrical patterns.

SLIDES

To make a slide pattern, slide copies of a shape from:

side to side up and down or across.

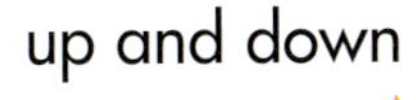

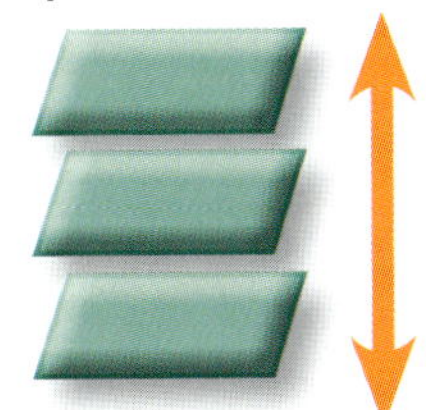

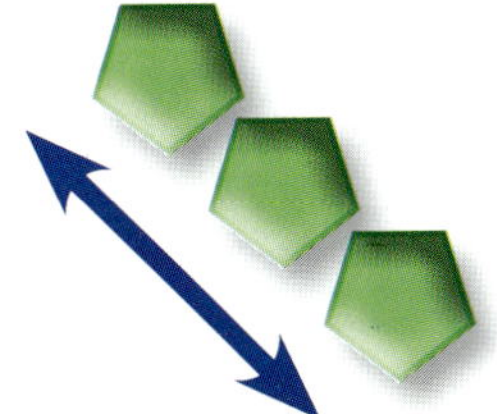

You can slide any shape.

Some slides look the same as a flip.

Many slides don't create symmetrical patterns.

TURNS

A turn pattern copies and rotates a shape around a centre point.

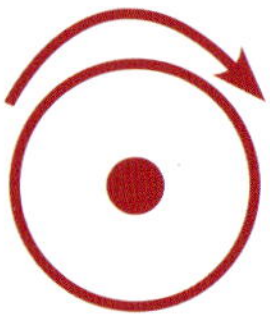

This is the **turning point**.

You can turn a shape **clockwise**.

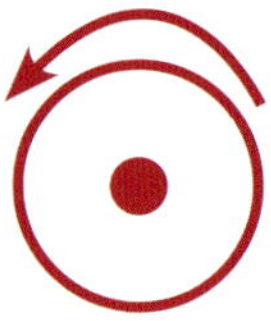

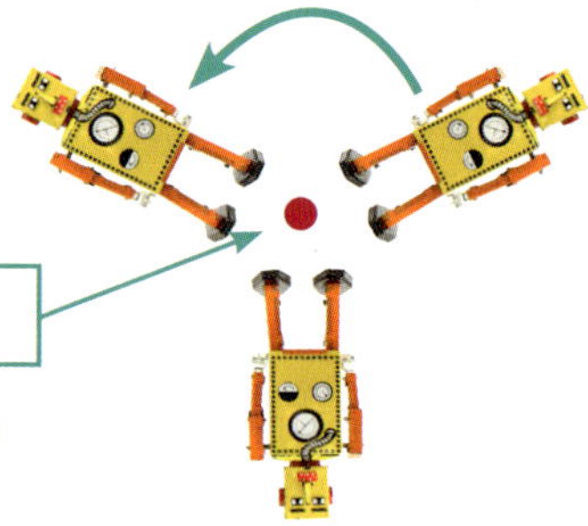

You can turn a shape **anti-clockwise**.

turning point

You can turn half a circle each time.

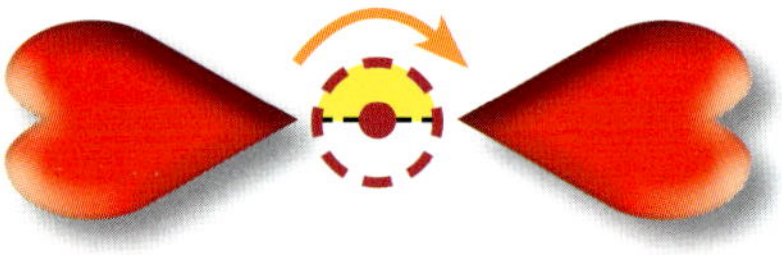

a half turn

This creates the same pattern as a flip.

quarter turns

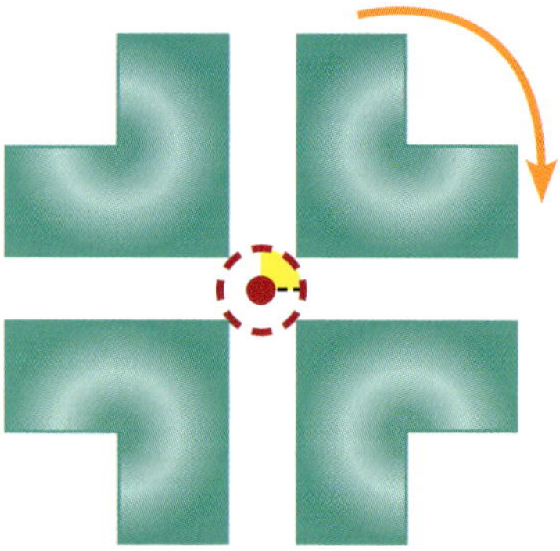

You can turn a quarter circle each time.

This creates the same symmetrical pattern as a flip.

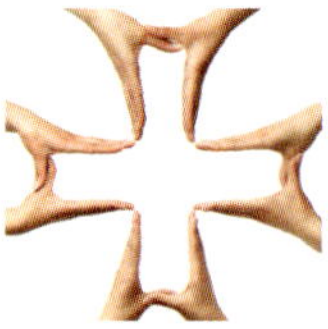

Try this

Look at each picture or pattern.

Write flip, slide or turn.

1.

2.

3.

4.

Location

POSITION WORDS

Every day you are somewhere.

The things around you are somewhere.

Position words help you describe a **location**.

at the park

backwards **in front of** **up** **forwards** **far** **above** **down** **around** **to the left** **near** **third** **below** **to the right** **in the middle of** **between**

Here are 3 things on top of Jack's desk.

Use different position words to describe their position to a friend.

FOLLOW DIRECTIONS

You use position words to give directions.

Where do you start?

Do you go forwards or backwards?

How far do you go?

Do you need to make a half or a quarter turn?

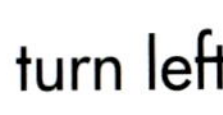

turn left

turn right

Give directions like '5 steps forwards', 'turn left at the shop' or 'go around the tree then 10 steps to the right'.

behind the fence

Sara is 4th

FOLLOW DIRECTIONS (continued)

Look at Lucy's treasure map.

Follow her directions.

Start at
Turn right. Go forwards.
Make a ¼ turn left.
Go left again
Take the top path to the end.
Where are you?

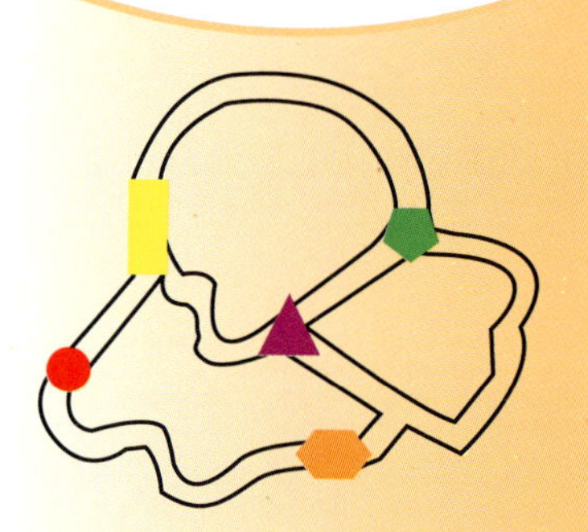

You are at the green pentagon.

Use Lucy's treasure map.

Write your own directions.

Ask a friend to find their way on the map.

You can give directions for copying a model.

Lee made this model.

She wrote directions to help you copy it.

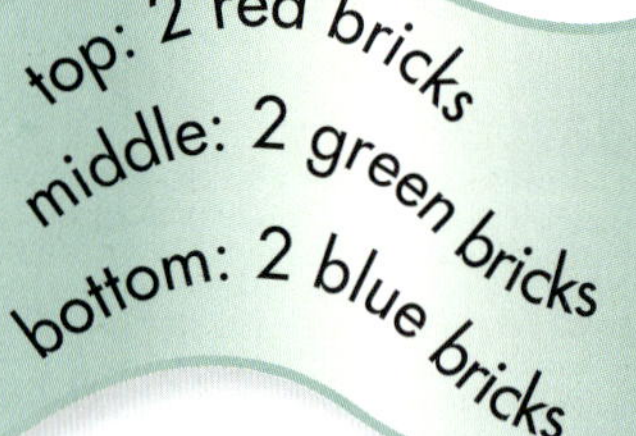

Challenge

Write directions for making Lee's shape picture.

MAKE A MODEL

A model is like a 3-dimensional map.

It shows the position of important objects.

Here are 3 models of this farmhouse.

You don't have to make it look real.

Here is a picture of a farm.

Your map doesn't have to be exact

trees

house

path

shed

garden

pond

You could draw it like this on a map

Use objects to make a model of your backyard.

Don't forget to label everything.

Challenge

Use position language to explain to a friend how to walk around your backyard.

USE A GRID

A grid helps you give better directions.

Look at these animals.

They are in 3 columns. Each column has a colour. 

They are in 3 rows. Each row has a number. **1 2 3**

The horse is in  **1**.

What animal is to the right of  **2**? The goat. It is in **2**.

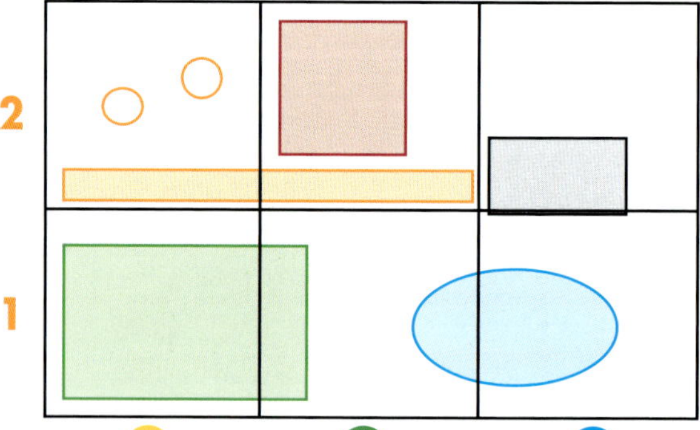

Imagine a grid like this on the farm map.

It can help you figure out where things are.

The farmhouse is in 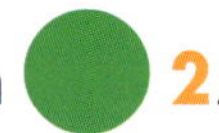 **2**.

Try this

Write 3 more position questions about the animals in the grid.

Challenge

Make a grid to fit these cars. Label your grid.

Describe the position of 2 cars in your grid.

SKETCH A MAP

You can draw your own map to help you locate position. Imagine looking down from above.

Josh drew his car like this:

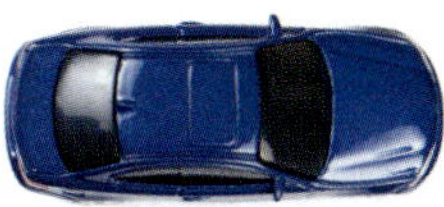

Gabi drew a map of her dinner plate like this:

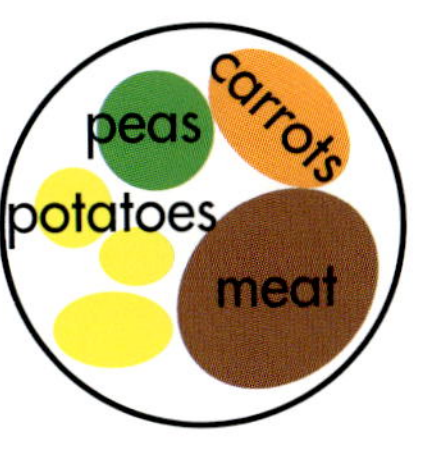

Ben made a sketch of his friends like this:

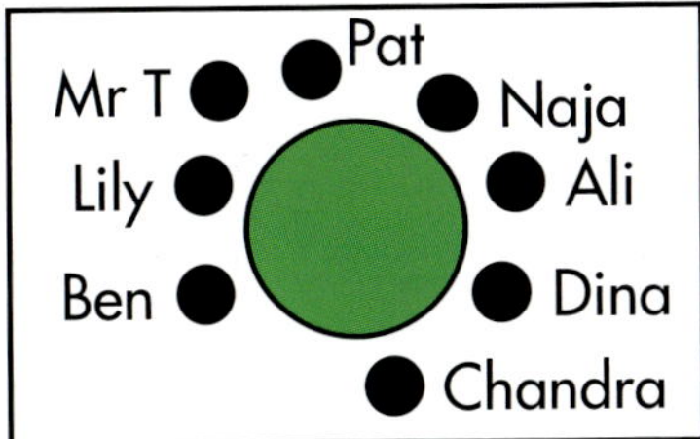

Ali is between Naja and Dina. Lily is opposite Dina.

Chandra is next to Ben and Dina.

Naja drew a map of tables and chairs in a café like this:

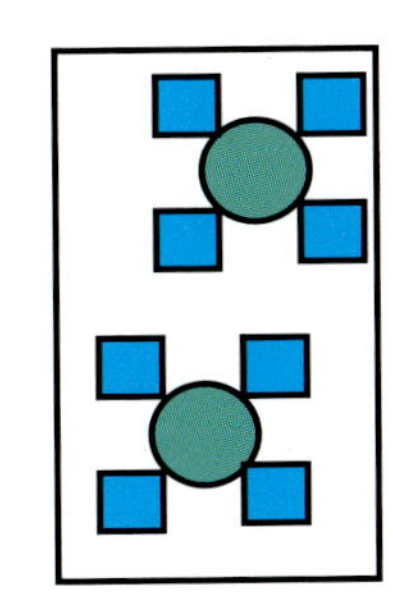

Try this

Sketch a map of the rooms inside your home.

Draw a path to show how to get from your bedroom to the kitchen.

Describe your map to a friend.

Chance is about whether something will or won't happen.

If you are sure of the result, it is **not** a chance event. It is **certain**.

- 1 week is 7 days.
- $1 is worth 100 cents.

If you are not sure of the result, it is a chance event. It is **uncertain**.

- You might have a sleepover on the weekend.
- You might have pasta for dinner.

Here are some words to help you talk about chance:

uncertain **possible** **might** **certain** **fair** **impossible** **unfair** **unlikely** **could** **likely**

POSSIBLE OR IMPOSSIBLE

Think about things that could happen.

They may not happen, but they could.

These events are all **possible**.

- Your pet bird learns to talk.

- Your family buys a new car.

- The spinner lands on yellow.

- You find 4 aces in a pack of cards.

Think about things that definitely won't happen.

These events are all **impossible**.

- Your pet fish learns to talk.

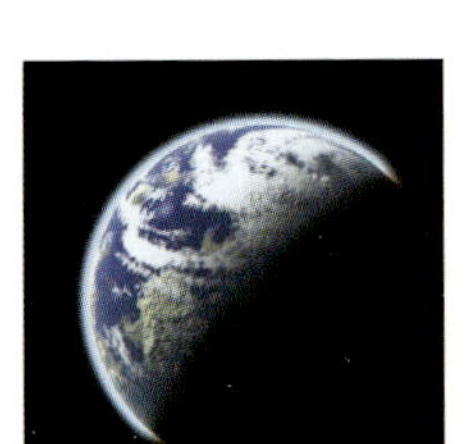

- Planet Earth spins backwards.

- April follows next after January.

- You grow younger every day.

LIKELY OR UNLIKELY

Think about things that might happen.

How likely are they to happen?

These events are all **likely** to happen.

- You eat breakfast in the morning.

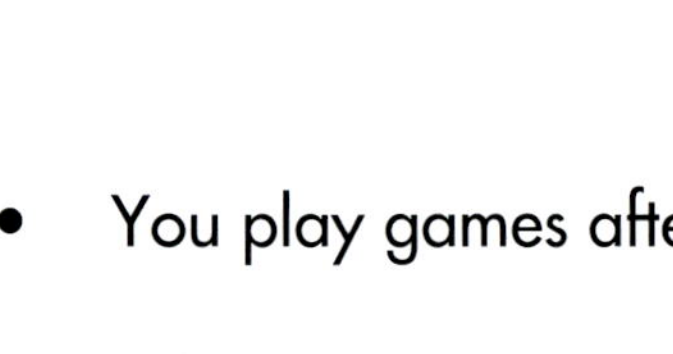

- You play games after school.

- You read a book this week.

You can't be sure they will happen.

They are not certain.

They are possible.

These events are all **unlikely** to happen.

- You win $5 in the lottery.

- You buy a green T-shirt today.

- Snow falls in summer.

They still could happen.

They are not impossible.

PREDICT A CHANCE EVENT

You can predict how a chance event will happen.
Your prediction won't always come true.

Ella predicts she will score 5 goals.
She scores 4 goals. 

Ryan predicts he will be in bed by 8 o'clock.
He was right. ☑

Molly predicts she will see a magpie.
She was right. ☑

People love to predict.

- Will a coin land on heads or tails?

- Will the ball you take out of the bag be red?

- Will the dice land on numbers less than 4?

You say you are lucky if your prediction is correct.
You say you are unlucky if your prediction is not correct.

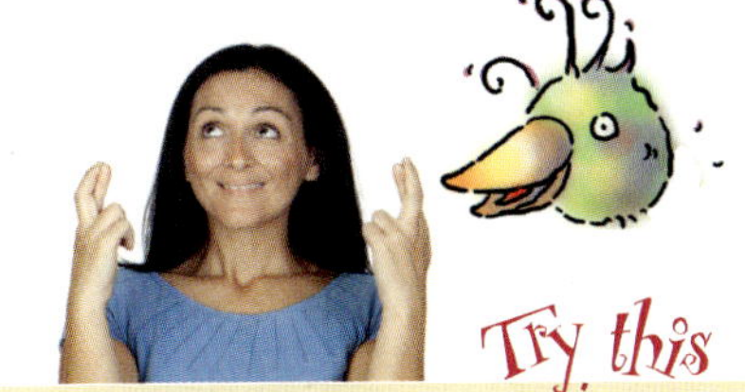

Match these chance labels to their event.

possible	Wednesday is 2 days after Monday
unlikely	You feel cold in winter
impossible	You visit your cousins this weekend
certain	You travel to Italy for a year
likely	Your dog reads the newspaper

DATA DISPLAYS

Data is about asking a useful question, then finding an answer to it.

What do you want to find out?

How many people will you ask?

What will you do with the answers you collect?

'YES OR NO' QUESTIONS

Think of a question that has 'yes' or 'no' as the answer.

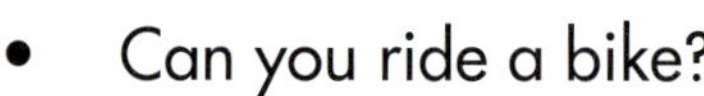

- Can you ride a bike?

- Are you usually in bed before 8:30 each night?

- Do you believe martians exist?
- Do you think you are good at art?

Decide how many people to ask.

The more people you ask the more data you collect.

Ceci wants to ask 10 people if they like to go to the beach.

She writes her question like this:

Ceci is now ready to ask 10 friends to answer her question.

Cheng wants to ask 15 friends this question:

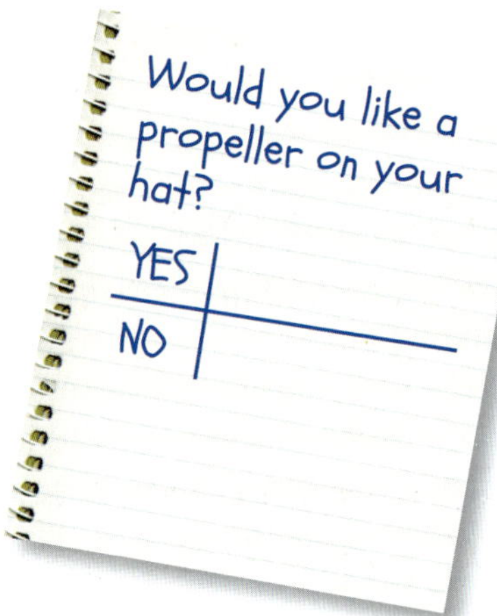

What 'yes or no' question would you ask?

'EITHER/OR' QUESTIONS

Think of a question with 2 alternatives.

- Do you spin a hula hoop better clockwise or anti-clockwise?
- Do you prefer peanut butter or Vegemite?
- Do you like your toast cut into halves or quarters?

Mimi wants to ask 20 friends if they prefer listening to music or stories.

She writes her question like this:

Mimi is ready to find out the answer.

KEEP A TALLY

Think about how to record everyone's answers.

A finger **tally** might get confusing.

Tally marks on paper are more useful.

This is what Ceci discovered:

Most people like the beach.

This is what Mimi discovered:

4 more of Mimi's friends liked stories.

This is what Cheng discovered:

Most of his friends don't want a hat propeller.

DISPLAY YOUR RESULTS

Use blocks and counters to display your results as a **graph**.

Ceci used shells to display her results.

Do you like to go to the beach?		
	Yes	7 shells
	No	3 shells

Mimi used buttons to display her results.

Do you like listening to stories or music?		
	Stories	12 buttons
	Music	8 buttons

Cheng used blocks to display his results.

Do you want a propeller on your hat?		
	Yes	6 blocks
	No	9 blocks

Remember to give your graph a name or title.

TALK ABOUT DATA

Use words like these to talk about your data displays.

fewer **the largest number of** **most** **more than half** **the smallest number of** **less than half**

Here are some things people discovered by asking questions:

- China has the most people in the world.
- Australians eat more ice-cream each than any other people in the world.
- There are more cows in India than in China.

ONE-TO-ONE PICTURE GRAPHS

You can use a **picture graph** to display your results.
Draw or cut out a picture for every fact you collect.

Your pictures can be horizontal 

Your pictures can be or vertical 

Ceci makes this graph from cut out pictures.
She puts for 'yes' and for 'no'.
She keeps the rows and columns tidy.

Do you like to go to the beach?

Yes	
No	

What else can Ceci say about her results?
Fewer people said 'no'.
More than half her friends like the beach.

COLUMN GRAPHS

A **column graph** uses a space on a grid to represent each fact you collect.

It is easier to colour in a space than to draw or cut out a picture.

You can even just put a cross in the space. X

The columns can be horizontal → or vertical ↑.

The top of each column matches a number on the left.

Chad asks 20 friends about hula-hoop preferences.

This is what he discovers.

How we spin a hoop

Number of children: 10, 9, 8, 7, 6, 5, 4, 3, 2, 1

clockwise | anti-clockwise | can't spin

Type of spin

I can't spin a hoop

Chad puts a label under the columns.

He puts a label along the side of the rows.

He remembers to give his graph a title.

What can Chad learn from looking at his graph?

Most people spin a hula-hoop anti-clockwise.

7 people can't spin it at all.

Only 4 people spin it clockwise.

This is what Chad's graph looks like with the grid lines removed.

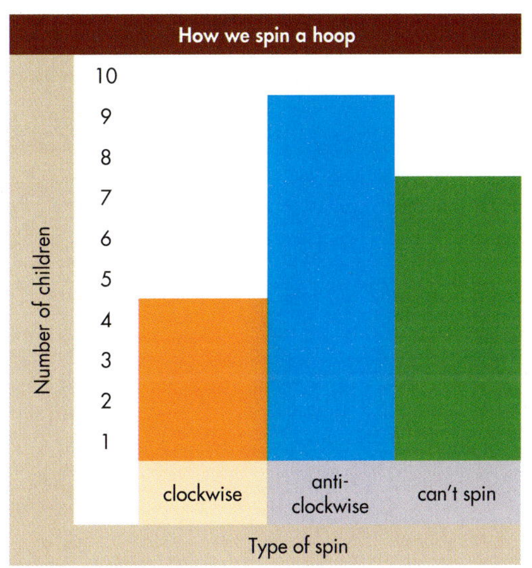

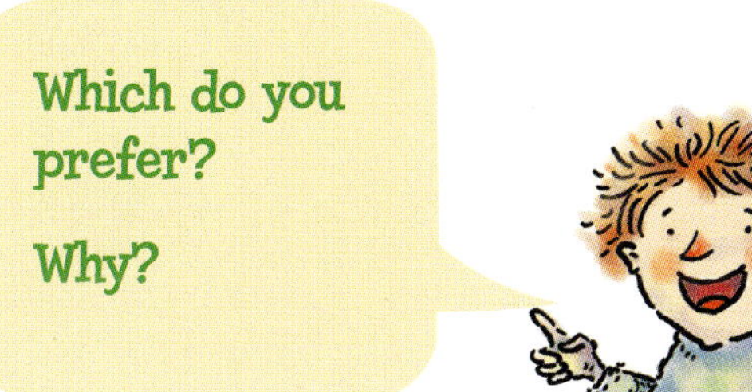

Look at this column graph.

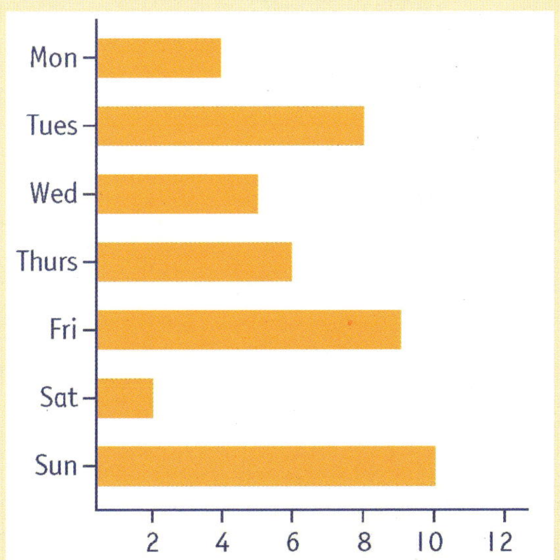

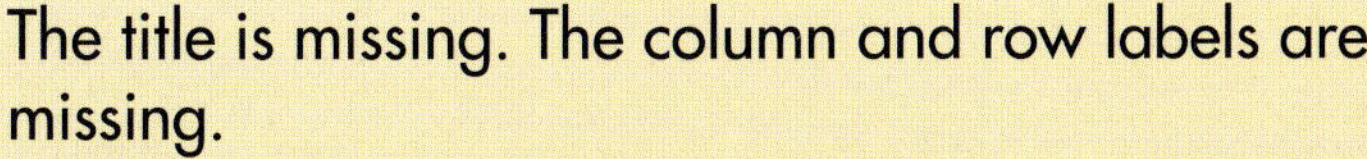

The title is missing. The column and row labels are missing.

What could this be about? Write your own title and labels.

Explain to a friend what your graph is about.

SYMBOLS & ABBREVIATIONS

Number

= is the same number as, is equal to

+ and, add

– take away, minus

× rows of, columns of, groups of

÷ shared between, put into equal groups of

Number	Word
10	ten
20	twenty
30	thirty
40	forty
50	fifty
60	sixty
70	seventy
80	eighty
90	ninety
100	one hundred

0	1	2	3	4	5	6	7	8	9
10	11	12	13	14	15	16	17	18	19
20	21	22	23	24	25	26	27	28	29
30	31	32	33	34	35	36	37	38	39
40	41	42	43	44	45	46	47	48	49
50	51	52	53	54	55	56	57	58	59
60	61	62	63	64	65	66	67	68	69
70	71	72	73	74	75	76	77	78	79
80	81	82	83	84	85	86	87	88	89
90	91	92	93	94	95	96	97	98	99

Number	Word
100	hundred
200	two hundred
300	three hundred
400	four hundred
500	five hundred
600	six hundred
700	seven hundred
800	eight hundred
900	nine hundred
1 000	one thousand

100 1 hundred
200 2 hundred
300 3 hundred
400 4 hundred
500 5 hundred
600 6 hundred
700 7 hundred
800 8 hundred
900 9 hundred
1000 10 hundred

Fractions

1 whole							
1 half				1 half			
1 quarter		1 quarter		1 quarter		1 quarter	
1 eighth	1 eighth	1 eighth	1 eighth	1 eighth	1 eighth	1 eighth	1 eighth

Measurement: Length

cm centimetre

m metre

Space

2D two-dimensional

3D three-dimensional

SELECTED 'TRY THIS' AND 'CHALLENGE' ANSWERS

NUMBER & ALGEBRA

Page 1 Count to 100

The girl in the brown suit is 9th.

Page 10 Count to 100

0	1	2	3	4	5	6	7	8	9
10	11	12	13	14	15	16	17	18	19
20	21	22	23	24	25	26	27	28	29
30	31	32	33	34	35	36	37	38	39
40	41	42	43	44	45	46	47	48	49
50	51	52	53	54	55	56	57	58	59
60	61	62	63	64	65	66	67	68	69
70	71	72	73	74	75	76	77	78	79
80	81	82	83	84	85	86	87	88	89
90	91	92	93	94	95	96	97	98	99

Answers will vary. Columns have the same 1s digit, rows have the same 10s digit. You add 11 if you go diagonally.

Page 21 Tens and ones

64

10s	1s
6	4

Page 25 Round to the nearest 10

1 40
2 70
3 100
4 50

Page 31 Rearranging 3-digit numbers

907
079
709 or 790

Page 37 Add to 10

4, 6, 3, 8

Page 39 Addition words and symbols

Stories will vary about 2 + 7 = 9
e.g. Lee has 2 frogs. Her friend Sara gives her 7 more frogs. Lee now has 9 frogs.

Page 43 Take away from 10

1 9 take away 4 leaves 5

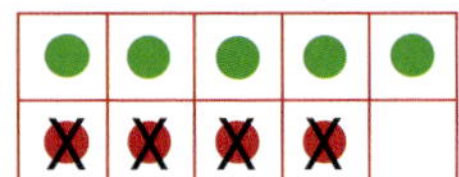

2

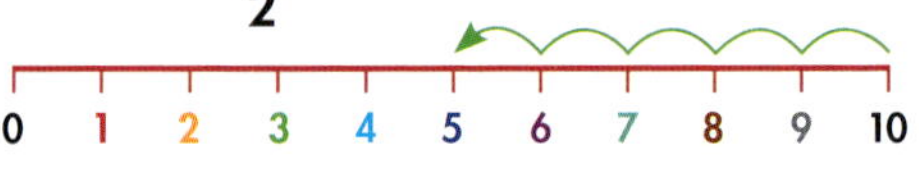

Page 44 Count back

8 take away **5** leaves 3
9 take away **7** leaves 2

Page 46 Number fact families

Answers will vary.

Page 51 Facts to 20 chart

Answers will vary e.g. 10 + 4 =14, 14 − 5 = 9, 6 + 8 = 14, 14 − 7 = 7.

Page 54 Record on a number line

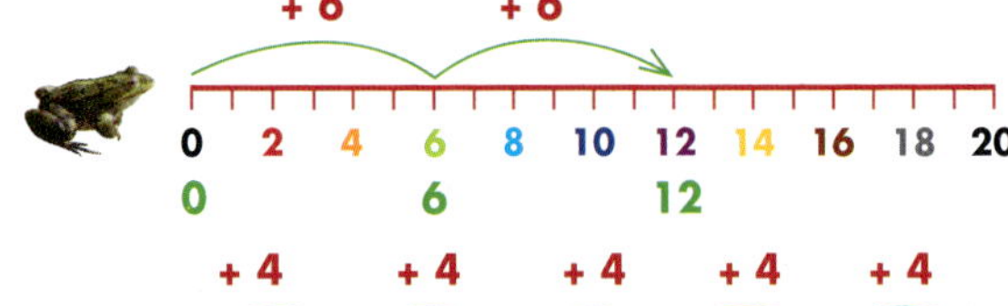

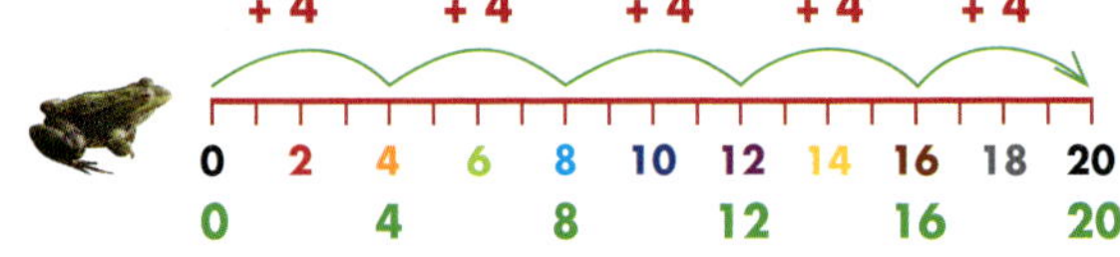

Page 55 Use an array

8 × 3 = 3 + 3 + 3 + 3 + 3 + 3 + 3 + 3 = 24

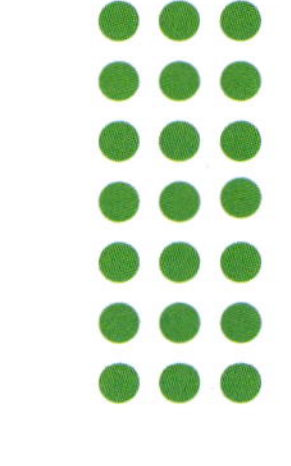

Page 57 Groups of 2

5 × 2 = 10

Page 59 Groups of 5

1 5 × 5 = 25
2 5 × 5 = 25
3 ⟶

Page 61 **Challenge**

Answers vary. e.g. 8 cats each had 5 kittens. That's 5 + 5 + 5 + 5 + 5 + 5 + 5 + 5 = 40 kittens.

Page 67 **More examples of groups and shares**

4 groups of 3 beetles with 1 extra

Page 81 **Combining coins**

Answers vary e.g. \$1, \$2, 50c, 20c and 5c

Or 4 × 50c, \$1, 3 × 20c, 10c and 5c

Challenge

3 pencils cost \$2 + \$2 + \$2 = \$6

2 books cost \$9 + \$9 = \$18

\$6 + \$18 = \$24

You can add \$6 to make \$30 and another \$20 to make \$50.

She will get \$26 change.

Page 86 **Plastic notes**

Answers vary e.g. \$50 or \$20 + \$20 + \$10 or 4 × \$5 and \$20 and \$10.

Challenge

Answers vary e.g. 2 × \$25, 3 × \$10, 1 × \$20

Page 88 **Make a pattern with objects**

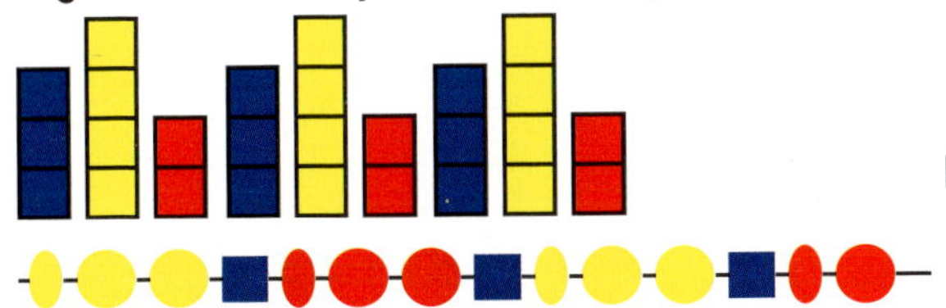

Page 90 **Add a number patterns**

48, 58, 68

30, 36, 42

Page 92 **Subtract a number patterns**

56 51 **46** 41 36 **31**

62 55 48 41 **34** 27

Challenge

Answers will vary, e.g.

The number above is 10 less.

The number below is 10 more.

The 1s digits in a column are the same.

The 10s digits in a column increase by 1.

MEASUREMENT & GEOMETRY

Page 95 **Compare curved or crooked lengths**

Answers will vary.

Page 97 **Measure with length units**

Measurements will vary.

Challenge

Each footstep measurer will be unique.

Page 99 **Measure with metres**

Body metres will all measure exactly 1 m but will vary in how far they reach along each person's arm.

Page 101 **Measure with centimetres**

The man has not lined the head of the fish at 0 on the left of his ruler.

The tail is sticking out longer than the ruler.

Page 103 **Comparing 2 areas**

Explanations will vary.

Page 107 **Measuring area with units**

Areas will vary. Trace your body outline onto a large sheet of paper. How many book units cover this area? Trace the shape of a book onto your body outline until it is covered. Count how many units you needed. Is this number more than you needed to cover the desk area? The larger the number of units the larger the area.

Page 109 **Measure area with a grid**

9 square units cover one face of this cube. A cube has 6 faces. That's 9 + 9 + 9 + 9 + 9 + 9 = 54 square units of area.

Page 111 **Comparing mass**

Answers will vary. It is very difficult to feel the difference when you heft small masses. Try comparing and hefting larger objects.

Page 114 **Put masses in order**

The toy dog is heavier than the car. So it is heavier than 18 shells.

It is lighter than the rabbit.

It is lighter than 21 shells.

It must be as heavy as 19 or 20 shells.

Page 123 **Read time on the hour**

6:00 11:00 8:00 4:00

Page 124 **Half past**

Gabi cleans her teeth at **8 o'clock**.

She goes to sleep at **30 minutes past 8**.

That's **8:30** at night. 8:30

Page 125 A quarter past

Oscar takes 15 minutes to eat his yoghurt.
He starts at **5 o'clock**.
He finishes at **a quarter past 5**.
That's **5:15**.

5:15

Page 127 A quarter to

Abba is checked by the vet from 5:00 to **5:45**.
That's 45 minutes.

She finishes at a **quarter to 6**.

Page 128 Days of the week

Monday: "moon day" (named after the moon)
Tuesday: "Tyr's day" (named after the old Scandinavian god of war, Tyr)
Wednesday: "Wodin's day" (named after the old Scandinavian god Wodin)
Thursday: "Thor's day" (named after the old Scandinavian god of thunder, Thor)
Friday: "Fria's day" (named after the old Scandinavian god of love, Freya)
Saturday: "Saturn's day" (named after the planet Saturn)
Sunday: "Sun's day" (named after the star we call the sun)

Challenge

This sign shows when Sally's shop is open.
Tick the times below when you can buy flowers.

- ☒ Mon 9:00
- ☑ Tues 11:30
- ☒ Wed 3:00
- ☑ Thurs 10:45
- ☑ Fri 9:15
- ☑ Sat 12:45
- ☒ Sun 4:00

Page 131 Read a calendar

Chinese New Year varies from year to year.
2013 = 10 February
2014 = 31 January
2015 = 19 February.

Page 132 Read a calendar

Challenge

Answers will vary.

SPACE

Page 142 Cones

Descriptions will vary.

1. Cone
2. Hexagonal prism
3. Sphere
4. A very thin cylinder
5. Triangular pyramid
6. Rectangular prism

Page 152 Tangrams

1 A quadrilateral

2 A pentagon

3 A hexagon

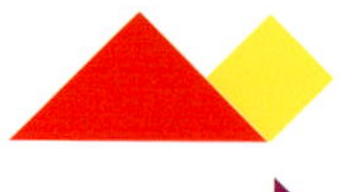

4 An octagon

Hidden shapes

1

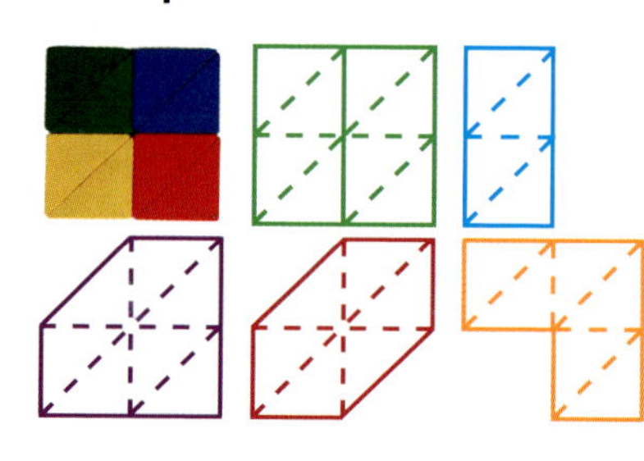

2

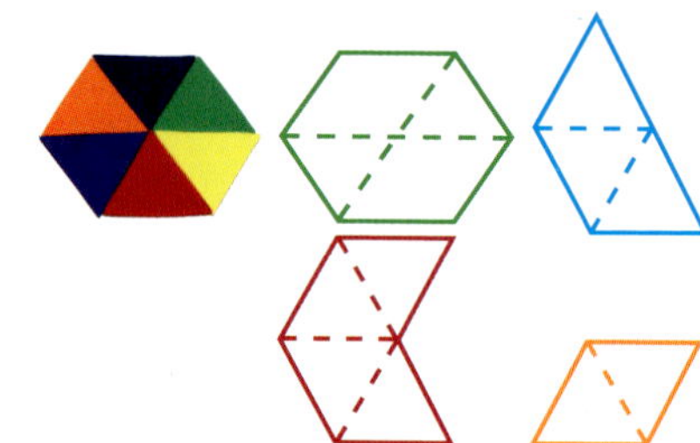

Page 154 Turns

1 6 turns

2 A flip across a vertical line of symmetry

3 4 possible flips (or 4 turns)

4 Slide pattern

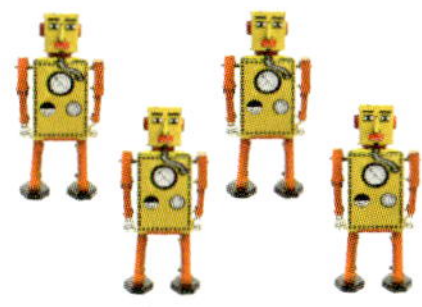

Page 155 Position words

Answers will vary. The cup is to the left of the pencils. The cup is 1st. The pencils are in the middle, between the cup and the robot. The pencils are 2nd. The robot is on the right of the pencils. It is 3rd.

Page 156 Follow directions

Map directions will vary.

Challenge

Directions will vary, e.g.
Join 2 triangles to make a square sail.
Make the boat from 2 squares side by side with a triangle at each end.
It is now a quadrilateral.

Page 158 Use a grid

Answers will vary, e.g.
Where is the horse? ● 1

Challenge

Answers will vary, e.g.

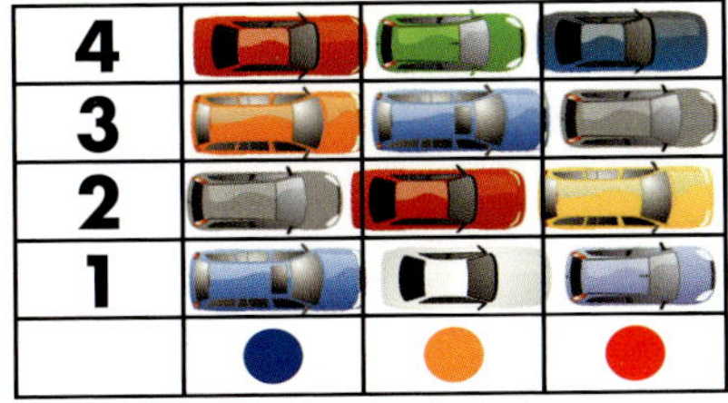

The orange car is in ● 3.
The green car is in ● 4.

STATISTICS & PROBABILITY

Page 163 Predict a chance event

Certain: Wednesday is 2 days after Monday.
Likely: You feel cold in winter.
Possible: You visit your cousins this weekend.
Unlikely: You travel to Italy for a year.
Impossible: Your dog reads the newspaper.

Page 169 Column graphs

Answers will vary, e.g.
The number of phone calls your family makes in a week
People rescued by lifesavers in a week at your favourite beach

GLOSSARY

Words that you might need to know.

addition	The operation of combining two or more numbers to find the total
analogue clock	A clock or watch with a circular face divided into 12 equal parts, showing time as the angle between two rotating hands
anti-clockwise	Circular movement in the opposite direction to the hands of an analogue clock
apex	The highest point, or pointy top, of something such as a triangle, a cone or a pyramid
area	The amount of surface a shape covers. It is measured in square units such as square metres
array	A grid system of rows and columns for sorting objects into equal groups
balance scales	An instrument used to measure mass. Objects placed on each side of the scales will balance if the objects are equal in mass
calendar	All the months of the year in order with the weeks in rows and the days in columns
capacity	A measure of how much a container can hold, or how much volume is inside a container
chance	The possibility of something happening
circle	A special 2D shape with only 1 curved side and no corners
clockwise	Circular movement in the same direction as the hands of an analogue clock
column graph	A graph that displays data using separate columns to show how many times something happens
cone	A 3D solid with 1 circular base and 1 curved surface that ends at a point. It looks like a circular pyramid
corner	A place where 2 or more sides of a 2D shape, or 2 or more edges of a 3D object, meet
cube	A special rectangular prism with 6 square faces, 8 corners and 12 edges
cubical	A 3D object that looks like a cube
curved surface	The outside layer of the curved part of a 3D object such as a cylinder, a cone or a sphere

cylinder	A 3D object with 2 identical circular faces at either end of one curved surface
cylindrical	A way to describe any 3D object that looks like a cylinder
data	Information that has been collected
data display	A graph or chart to show the information collected
date	A system for measuring time in days, months and years
difference	The result in a subtraction problem when you find the amount by which one number is larger or smaller than another
digital clock	A device for measuring and displaying time as numbers from 00:00 to 11:59
digits	The ten symbols from 0 to 9 used to record numbers in groups of 10
division	The operation of putting things into equal size groups by sharing or by grouping using repeated subtraction
edge	The line on a 3D object where 2 faces meet
even number	A whole number that can be put into pairs or groups of 2 with no left overs
face	Any flat 2D surface on a 3D object
fair share	Sort or divide things into equal size groups
flip	Make a mirror image by reflecting a 2D shape across a line of symmetry
gram	A small metric unit for measuring the heaviness or mass of an object
grid map	A map with horizontal (going across) and vertical (up and down) lines that can be used to locate a position
half	One of two equal parts
heft	To estimate the heaviness or mass of an object by holding it in your hands
horizontal lines	Straight lines that go from left to right like the horizon and are parallel to the ground
hour	A unit of time where one day is divided into 2 groups of 12 (or 24) equal parts from midnight onwards until the next midnight
hundreds	A 3-digit number representing groups of 10 × 10 in our counting system, including any number from 101 to 999
hundreds place	The number in the third position from the right in any whole number equal to or larger than 100
Leap Year	Every 4 years an extra day is added to February to give 366 days

left over	The amount that sometimes remains when a group of objects is shared into equal groups
line of symmetry	A mirror line where 2 reflected halves meet
litre	A metric unit for measuring volume and capacity
location	Your position in the world around you or the position of objects
mass	A measure of heaviness or the amount of matter in an object
metre	The standard unit for measuring length. It is as long as 4 Maths Guides
midday	The middle of the day or 12:00, sometimes called noon
minus	Another name for take-away or subtract, where you find the difference between 2 numbers
minute	A small unit of time where one hour is divided into 60 equal parts
mirror line	The line where 2 halves meet on a symmetrical shape
month	One of 12 parts of a year.
multiplication	The operation of repeatedly adding the same number to get a product
oblong	A rectangle with 2 shorter parallel sides, 2 longer
odd number	A whole number that has one left over when you put many objects into pairs or groups of 2
ones	Another name for units in our counting system. Also the name for whole numbers smaller than 10, or any numbers left over when counting out a group of 10
ones place	The number in the first position from the right in any whole number
parallel lines	2 or more lines of any length that never touch each other as they are the same distance apart
picture graph	A graph that uses pictures or symbols to show each item of data in a collection
place value	The value of where a digit is in a number, such as thousands, hundreds, tens or ones. In the number 3496 the 3 means 3 thousands
polygon	A 2D shape with 3 or more straight sides. Some special polygons are pentagons (5 sides), hexagons (6 sides) and octagons (8 sides)
prism	A 3D object with 2 identical flat faces or bases at either end, joined by 3 or more rectangular faces

pyramid	A 3D object with one base and 3 or more triangular faces that meet at a point
rectangle	A 2D shape with 4 straight sides and 4 square corners
reflection	A mirror image
rim	The top of a container
slide	Move a shape to the right or left, or up or down
spherical	A 3D object that looks like a ball or a sphere
square	A rectangle with 4 sides of equal length and 4 square corners
surface	The flat area inside a 2D shape or the outside area of a 3D object
symmetrical	Any shape which has at least one mirror line or line of symmetry, where one half is perfectly reflected in the other half
tangram	A puzzle in the shape of a square broken into 7 pieces that can be rearranged to make new shapes
tens	A 2-digit number in our counting system, representing groups of 10, including any number from 10 to 99
tessellate	To cover an area with 2D shapes so that there are no gaps and no overlaps
thousands	A 4-digit number in our counting system, representing groups of 10 × 10 × 10, including any number from 1000 to 9999
three-dimensional or 3D	Objects that have height, width and depth
transformation	The result of moving a shape. A slide, a flip or a turn
triangle	Any polygon with 3 straight sides
turn	The result of moving a shape by rotation clockwise or anti-clockwise
two-dimensional or 2D	Flat shapes that have both height and width
unfair share	Sort or divide things into groups of different sizes
vertical lines	Straight lines in an up and down position
volume	A measure of how much space a 3D object takes up
week	7 days. There are 52 weeks in one year
weekdays	The 5 days from Monday to Friday in each week.
weekend	The 2 days from Saturday to Sunday in each week
year	365 days or 366 days in a leap year when February has an extra day

INDEX